Detecting and Responding to Alien

Ecologists, land managers, and policy makers continue to search for the most effective ways to manage biological invasions. An emerging lesson is that proactive management can limit negative impacts, reduce risks, and save money. This book explores how to detect and respond to alien plant incursions, summarising the most current literature, providing practical recommendations, and reviewing the conditions and processes necessary to achieve prevention, eradication, and containment. Chapter topics include assessing invasiveness and the impact of alien plants, how to improve surveillance efforts, how to make timely management decisions, and how legislation and strategic planning can support management. Each chapter includes text boxes written by international experts that discuss topical issues such as spatial predictive modelling, costing invasions, biosecurity, biofuels, and dealing with conflict species.

JOHN R. WILSON has published close to one hundred papers in peer-reviewed journals on a wide range of ecological and evolutionary topics, with a particular focus on invasion science. Based in South Africa, he is a member of the IUCN Invasive Species Specialist Group, and works across science, management, and policy.

F. DANE PANETTA has over forty years of weeds-related research experience in the areas of ecology, risk assessment, and incursion management. He has published close to one hundred papers in peer-reviewed journals and has edited or co-edited three books on weed risk assessment. He resides in Brisbane, Australia.

CORY LINDGREN is an ecologist and biosecurity policy analyst who earned his PhD from the University of Manitoba. He is a long time director of the Invasive Species Council of Manitoba and on the organizing committee for the Weeds Across Borders conference. He has authored numerous peer-reviewed papers and contributed chapters to books on invasive species.

ECOLOGY, BIODIVERSITY, AND CONSERVATION

Series Editors
Michael Usher   *University of Stirling, and formerly Scottish Natural Heritage*
Denis Saunders   *Formerly CSIRO Division of Sustainable Ecosystems, Canberra*
Robert Peet   *University of North Carolina, Chapel Hill*
Andrew Dobson   *Princeton University*

Editorial Board
Paul Adam   *University of New South Wales, Australia*
H. J. B. Birks   *University of Bergen, Norway*
Lena Gustafsson   *Swedish University of Agricultural Science*
Jeff McNeely   *International Union for the Conservation of Nature*
R. T. Paine   *University of Washington*
David Richardson   *University of Stellenbosch*
Jeremy Wilson   *Royal Society for the Protection of Birds*

The world's biological diversity faces unprecedented threats. The urgent challenge facing the concerned biologist is to understand ecological processes well enough to maintain their functioning in the face of the pressures resulting from human population growth. Those concerned with the conservation of biodiversity and with restoration also need to be acquainted with the political, social, historical, economic, and legal frameworks within which ecological and conservation practice must be developed. The new Ecology, Biodiversity, and Conservation series will present balanced, comprehensive, up-to-date, and critical reviews of selected topics within the sciences of ecology and conservation biology, both botanical and zoological, and both 'pure' and 'applied'. It is aimed at advanced final-year undergraduates, graduate students, researchers, and university teachers, as well as ecologists and conservationists in industry, government, and the voluntary sectors. The series encompasses a wide range of approaches and scales (spatial, temporal, and taxonomic), including quantitative, theoretical, population, community, ecosystem, landscape, historical, experimental, behavioural, and evolutionary studies. The emphasis is on science related to the real world of plants and animals rather than on purely theoretical abstractions and mathematical models. Books in this series will, wherever possible, consider issues from a broad perspective. Some books will challenge existing paradigms and present new ecological concepts, empirical or theoretical models and testable hypotheses. Other books will explore new approaches and present syntheses on topics of ecological importance.

Ecology and Control of Introduced Plants
Judith H. Myers and Dawn Bazely

Invertebrate Conservation and Agricultural Ecosystems
T. R. New

Risks and Decisions for Conservation and Environmental Management
Mark Burgman

Ecology of Populations
Esa Ranta, Per Lundberg and Veijo Kaitala

Nonequilibrium Ecology
Klaus Rohde

The Ecology of Phytoplankton
C. S. Reynolds

Systematic Conservation Planning
Chris Margules and Sahotra Sarkar

Large-Scale Landscape Experiments: Lessons from Tumut
David B. Lindenmayer

Assessing the Conservation Value of Freshwaters: An international perspective
Philip J. Boon and Catherine M. Pringle

Insect Species Conservation
T. R. New

Bird Conservation and Agriculture
Jeremy D. Wilson, Andrew D. Evans, and Philip V. Grice

Cave Biology: Life in darkness
Aldemaro Romero

Biodiversity in Environmental Assessment: Enhancing ecosystem services for human well-being
Roel Slootweg, Asha Rajvanshi, Vinod B. Mathur, and Arend Kolhoff

Mapping Species Distributions: Spatial inference and prediction
Janet Franklin

Decline and Recovery of the Island Fox: A case study for population recovery
Timothy J. Coonan, Catherin A. Schwemm, and David K. Garcelon

Ecosystem Functioning
Kurt Jax

Spatio-Temporal Heterogeneity: Concepts and analyses
Pierre R. L. Dutilleul

Parasites in Ecological Communities: From interactions to ecosystems
Melanie J. Hatcher and Alison M. Dunn

Zoo Conservation Biology
John E. Fa, Stephan M. Funk, and Donnamarie O'Connell

Marine Protected Areas: A multidisciplinary approach
Joachim Claudet

Biodiversity in Dead Wood
Jogeir N. Stokland, Juha Siitonen, and Bengt Gunnar Jonsson

Landslide Ecology
Lawrence R. Walker and Aaron B. Shiels

Nature's Wealth: The economics of ecosystem services and poverty
Pieter J. H. van Beukering, Elissaios Papyrakis, Jetske Bouma, and Roy Brouwer

Birds and Climate Change: Impacts and conservation responses
James W. Pearce-Higgins, and Rhys E. Green

Marine Ecosystems: Human impacts on biodiversity, functioning and services
Tasman P. Crowe and Christopher L. J. Frid

Wood Ant Ecology and Conservation
Jenni A. Stockan and Elva J. H. Robinson

Detecting and Responding to Alien Plant Incursions

JOHN R. WILSON
Principal Scientist, South African National Biodiversity Institute, Kirstenbosch Research Centre, Claremont, South Africa, and Core Team Member, Centre for Invasion Biology, Department of Botany and Zoology, Stellenbosch University, Stellenbosch, South Africa

F. DANE PANETTA
Honorary Principal Fellow, Faculty of Veterinary and Agricultural Sciences, The University of Melbourne, Parkville, Victoria, Australia

CORY LINDGREN
Invasive Species Specialist, Manitoba, Canada

CAMBRIDGE
UNIVERSITY PRESS

University Printing House, Cambridge CB2 8BS, United Kingdom

Cambridge University Press is part of the University of Cambridge.

It furthers the University's mission by disseminating knowledge in the pursuit of education, learning and research at the highest international levels of excellence.

www.cambridge.org
Information on this title: www.cambridge.org/9781107095601

© Cambridge University Press 2017

This publication is in copyright. Subject to statutory exception and to the provisions of relevant collective licensing agreements, no reproduction of any part may take place without the written permission of Cambridge University Press.

First published 2017

Printed in the United Kingdom by TJ International Ltd. Padstow Cornwall

A catalogue record for this publication is available from the British Library

Library of Congress Cataloguing in Publication data
Names: Wilson, John R., 1977– author.
Title: Detecting and responding to alien plant incursions / John R. Wilson, F. Dane Panetta, Cory Lindgren.
Other titles: Ecology, biodiversity, and conservation.
Description: [New York] : Cambridge University Press, 2017. | Series: Ecology, biodiversity and conservation | Includes bibliographical references and index.
Identifiers: LCCN 2016019833 | ISBN 9781107095601
Subjects: LCSH: Invasive plants – Control. | Alien plants – Control.
Classification: LCC SB613.5 .W55 2016 | DDC 581.6/2 – dc23

LC record available at https://lccn.loc.gov/2016019833

ISBN 978-1-107-09560-1 Hardback
ISBN 978-1-107-47948-7 Paperback

Cambridge University Press has no responsibility for the persistence or accuracy of URLs for external or third-party internet websites referred to in this publication, and does not guarantee that any content on such websites is, or will remain, accurate or appropriate.

Contents

List of Contributing Authors	page ix
Foreword	xii
Preface	xv
List of abbreviations	xviii

1 Introduction   1

    Box 1.1 Incursion Response in New Zealand – *Philip E. Hulme*   7

2 Prediction (Pre- and Post-Border)   19

    Box 2.1 Plant Traits Associated with Impact on Native Plant Species Richness – *Montserrat Vilà, Rudolf P. Rohr, José L. Espinar, Philip E. Hulme, Jan Pergl, J. Jacobus Le Roux, Urs Schaffner, & Petr Pyšek*   29

    Box 2.2 Lag Phases: Theory, Data, and Practical Implications – *Petr Pyšek*   33

    Box 2.3 Species Distribution Models – *Jane Elith*   41

3 Detection and Delimitation   52

    Box 3.1 Risk Mapping to Underpin Post-Border Weed Management Activities – *Rieks D. van Klinken & Justine V. Murray*   59

    Box 3.2 Estimating Detectability Using Search Experiments – *Cindy E. Hauser & Joslin L. Moore*   71

4 Evaluation of Management Options   80

    Box 4.1 Is it Feasible to Eradicate or Contain Plant Incursions in the Galapagos Islands? – *Mark R. Gardener*   96

5 Evaluation of Management Performance	**111**
Box 5.1 Allocating Resources – *Oscar Cacho*	131
6 Legislation and Agreements	**139**
Box 6.1 Legislation in Antarctica – *Dana M. Bergstrom & Justine D. Shaw*	141
Box 6.2 Regulating the Use of Potential Invaders for Bioenergy – *Lauren D. Quinn*	161
Box 6.3 Managing Invasive Ornamental Trees: Conflicting Views and Values in Hawai'i – *Curtis C. Daehler*	164
7 Strategies and Actions	**169**
Box 7.1 National Strategies for Dealing with Biological Invasions: South Africa as an Example – *Brian W. van Wilgen*	170
Box 7.2 Costing Invasions in the UK – *Richard H. Shaw*	188
8 Implementation	**193**
Box 8.1 What is a Cooperative Weed Management Area? – *Al Tasker*	200
Box 8.2 The European and Mediterranean Plant Protection Organization: Coordinating the Response to Invasive Plants Across Borders – *Sarah Brunel*	201
Box 8.3 Invasive Species Early Detection and Rapid Response (EDRR): A Land Conservation Challenge for the Twenty-First Century – *Randy Westbrooks & Steven Manning*	207
Box 8.4 Raising Awareness About Invasive Plants in Portugal – *Elizabete Marchante & Hélia Marchante*	211
9 Conclusions and Future Directions	**217**
Glossary	228
References	238
Index	263

Contributing Authors

DANA M. BERGSTROM – BOX 6.1
Australian Antarctic Division, Kingston, Tasmania, Australia

SARAH BRUNEL – BOX 8.2
European and Mediterranean Plant Protection Organization (EPPO/OEPP), Paris, France

OSCAR CACHO – BOX 5.1
UNE Business School, University of New England, Armidale, New South Wales, Australia

CURTIS C. DAEHLER – BOX 6.3
Department of Botany, University of Hawai'i at Manoa, Honolulu, Hawai'i, USA

JANE ELITH – BOX 2.3
School of BioSciences, University of Melbourne, Parkville, Victoria, Australia

JOSÉ L. ESPINAR – BOX 2.1
Estación Biológica de Doñana (EBD-CSIC), Isla de la Cartuja, Sevilla, Spain

MARK R. GARDENER – BOX 4.1
School of Plant Biology, University of Western Australia, Crawley, Western Australia, Australia

CINDY E. HAUSER – BOX 3.2
School of BioSciences, University of Melbourne, Parkville, Victoria, Australia

PHILIP E. HULME – BOX 1.1, BOX 2.1
Bio-Protection Research Centre, Lincoln University, Lincoln, New Zealand

J. JACOBUS LE ROUX — BOX 2.1
Centre for Invasion Biology, Department of Botany and Zoology, Stellenbosch University, Stellenbosch, South Africa

STEVEN MANNING — BOX 8.3
Invasive Plant Control Inc., Nashville, Tennessee, USA

ELIZABETE MARCHANTE — BOX 8.4
Centro de Ecologia Funcional, Department of Life Sciences, University of Coimbra, Coimbra, Portugal

HÉLIA MARCHANTE — BOX 8.4
Centro de Ecologia Funcional, Department of Life Sciences, University of Coimbra, Coimbra, Portugal, and Coimbra Agriculture School, Polytechnic of Coimbra, Coimbra, Portugal

JOSLIN L. MOORE — BOX 3.2
School of Biological Sciences, Monash University, Melbourne, Victoria, Australia

JUSTINE V. MURRAY — BOX 3.1
CSIRO, EcoSciences Precinct, Brisbane, Queensland, Australia

JAN PERGL — BOX 2.1
Institute of Botany, The Czech Academy of Sciences, Průhonice, Czech Republic

PETR PYŠEK — BOX 2.1, BOX 2.2
Institute of Botany, The Czech Academy of Sciences, Průhonice, Czech Republic, and Department of Ecology, Faculty of Science, Charles University in Prague, Czech Republic

LAUREN D. QUINN — BOX 6.2
College of Agricultural, Consumer, and Environmental Sciences, University of Illinois, Urbana, Illinois, USA

RUDOLF P. ROHR — BOX 2.1
Unit of Ecology and Evolution, University of Fribourg, Fribourg, Switzerland

URS SCHAFFNER — BOX 2.1
CABI Switzerland, Delémont, Switzerland

JUSTINE D. SHAW — BOX 6.1
Australian Antarctic Division and School of Biological Sciences, University of Queensland, Queensland, Australia

RICHARD H. SHAW – BOX 7.2
CABI, Egham, Surrey, UK

AL TASKER – BOX 8.1
USDA-APHIS, Plant Protection and Quarantine (Retired), Philippi, West Virginia, USA

RIEKS D. VAN KLINKEN – BOX 3.1
CSIRO, EcoSciences Precinct, Brisbane, Queensland, Australia

BRIAN W. VAN WILGEN – BOX 7.1
Centre for Invasion Biology, Department of Botany and Zoology, Stellenbosch University, Stellenbosch, South Africa

MONTSERRAT VILÀ – BOX 2.1
Estación Biológica de Doñana (EBD-CSIC), Isla de la Cartuja, Sevilla, Spain

RANDY WESTBROOKS – BOX 8.3
Invasive Plant Control Inc., Whiteville, North Carolina, USA

Foreword

Invasion science has a relatively short history. Islands, as the parts of our planet most conspicuously invaded by non-native biota, gave rise to the first books on invasion science (Thomson 1922; Elton 1958). However, it was immediately clear that many continental areas had been heavily invaded as well. At the end of the last century, an organised and international biological invasions research effort (Simberloff 2011) centred around four basic questions:

(1) What makes some taxa more or less invasive?
(2) What makes some individual ecosystems more or less invasible?
(3) What are the impacts of invasive taxa?
(4) What should be done if impacts are economically and/or environmentally undesirable?

The usual assumption of the last question is that impacts have already been recognised. Indeed, attempts to control or even eradicate non-native weeds and pests have a long history in agriculture, forestry, and health care. More recently, such attempts were extended to protected areas such as nature reserves and national parks. However, with enormous investment into control efforts, it very soon became clear that introduction prevention and fast eradication after early detection of new invaders represent the most cost-effective and ecologically effective strategies. Still, decisions about whether and when recently introduced taxa should be monitored, contained, or eradicated remain a grey area. To date well over 100 books have been published on plant invasions. However, how to detect and how to respond to initial stages of plant invasions have not been systematically covered in any book publication until now.

For a long time, this 'knowing–doing' gap in invasion science has been recognised (Esler *et al*. 2010; Abella 2014; Matzek *et al*. 2015). This book, by John Wilson, Dane Panetta, and Cory Lindgren, substantially helps to close this gap. Because the book does not deal with specific invasions, it does not provide a cookbook for specific situations. Instead, using many examples, the reader is guided through the tools we have for (a) prediction of invasiveness and impacts; (b) detection and delimitation of new incursions of non-native plants; (c) eradication and/or containment feasibility; (d) evaluation of management results; (e) implementation and development of effective legislation and regulations; (f) prioritisation and development of feasible strategies; and (g) implementation of those strategies. To incorporate as much relevant knowledge as possible, 26 invited authors contributed text boxes summarising their respective expertise. These boxes provide perspectives and insights from across the globe, bringing the reader up to speed on the current state of invasion science. The Glossary and over 400 references will make this volume particularly useful.

Among books on biological invasions currently available on the market, this one is a unique achievement. I greatly look forward to using this book, and am confident that plant ecologists and managers around the world will find it both a valuable resource and a pleasure to read.

Marcel Rejmánek
University of California, Davis

References

Abella, S. R. (2014) Effectiveness of exotic plant treatments on National Park Service lands in the United States. *Invasive Plant Science and Management*, **7**, 147–163.

Elton, C. S. (1958) *The Ecology of Invasions of Animals and Plants*. Methuen, London.

Esler, K. J., Prozesky, H., Sharma, G. P., & McGeoch, M. (2010) How wide is the 'knowing–doing' gap in invasion biology? *Biological Invasions*, **12**, 4065–4075.

Matzek, V., Pujalet, M., & Cresci, S. (2015) What managers want from invasive species research versus what they get. *Conservation Letters*, **8**, 33–40.

Simberloff, D. (2011) SCOPE project. *Encyclopedia of Biological Invasions* (eds D. Simberloff & M. Rejmánek), pp. 617–619. University of California Press, Berkeley.

Thomson, G. M. (1922) *The Naturalization of Animals and Plants in New Zealand*. Cambridge University Press, Cambridge.

Preface

Alien plants can be managed most cost-effectively either before they have begun to spread or in their earliest stages of invasion, what we call in this book 'incursions'. When distributions are small in extent, it is possible, through coordinated control strategies, to achieve eradication or prevent further spread (i.e. containment). There has been an increasing focus on these management goals from a theoretical perspective and in relation to on-ground management, but much of the theoretical work either does not support current on-ground practice or fails to address the needs of managers and decision makers. A major aim of this book is to draw together the scattered literature and provide recommendations for how the management of incursions can be improved.

In this book we argue that incursions can be managed effectively, provided a few simple steps are taken. First, there is a need to predict, prevent, and prepare. This will help reduce the rate of new arrivals, and increase the rate at which incursions are detected. Second, when an incursion is detected, it is important not to wait, but to make an initial decision regarding whether or not to act. If the decision is to act, management options should be evaluated in a structured manner. Finally, for management to be effective it needs to be properly monitored so that performance can be evaluated and goals revisited if required. The success of an incursion response also depends on a host of facilitating activities and mechanisms – for example, legislation, strategies, action plans, and organisational structures.

The reader should not be overly concerned with the large number of alien plants that occur in different parts of the world. Most alien plant incursions pose a negligible threat and do not need to be managed (nor could they be, given the limitations on resources available). The first component of incursion response is deciding whether or not to act, and a proper response often will not involve on-ground management. A major motivation for writing this book has been to provide tools to help with this decision.

The book has been written for a range of potential readers, including land managers, policy makers, students, and researchers. While it is generally aimed at the science–policy interface, some parts review key emerging principles (e.g. assessing which species pose the greatest threat), while elsewhere we focus on outlining existing protocols (e.g. how to assess progress towards eradication). In some cases the discussion might therefore be too esoteric to be of direct value to land managers, and in other cases, where the state of knowledge is more advanced, the focus will be on how to do things in practice without discussing the fundamentals at length. We hope the reader will benefit from at least a cursory examination of all chapters, but as a rough guide: Chapters 2 and 3 address why and where plant invasions happen; Chapters 4 and 5 focus on deciding on the goal of an incursion response and how to monitor progress towards that goal; and Chapters 6, 7, and 8 discuss the policy and enabling environment in terms that often apply to all plant invasions.

Plant invasions are context-dependent – spatially, temporally, and socio-economically. Hence management needs to be adaptive. As such, we try to avoid being prescriptive and address the theories and management practices as we know them. We do, however, end each chapter with some general recommendations. Throughout the book international experts have contributed text boxes that elucidate particular topics of interest. These text boxes are vignettes that highlight real examples of what has, and has not, worked.

There are a couple of things that the book does not address. First, it is not going to 'make the case'. Invasive plants are a serious global threat to biodiversity, the environment, economies, trade, and human health. If you need convincing, there are some excellent reports available (e.g. Boy and Witt 2013; Box 7.2). Second, we elected not to discuss classical biological control as a management option. Classical biological control can provide substantial benefits in terms of reducing rates of spread and so should be used where and when practical. There is also much to be said about the proactive development and testing of agents. But, to date, biocontrol is only implemented once an incursion has become a widespread invasion. Finally, we do not discuss restoration. Invasions can sometimes be passengers rather than drivers of global change, and control efforts will only have long-term benefits if substantial effort is placed into restoring ecosystems, as otherwise the area treated might simply be recolonised by the same or another alien plant. However, the theory and practice of restoration is well discussed elsewhere. Moreover, if the impact of an invasion is such that restoration is required, the opportunity for coordinated control will have likely well and truly passed.

The book would, of course, not have been possible without the support of our friends, colleagues, and families. We are indebted to S. Raghu, Marcel Rejmánek, Dave Richardson, Brian van Wilgen, and Bruce Wilson for their comments and insights on earlier versions of the book; to the Invasives in the Cape Discussion Group, SANBI's Invasive Species Programme, and the past and present staff, students, and post-docs of the Centre for Invasion Biology for useful insights, enthusiasm, and debates; to the text box authors for providing much needed colour and poetry to the prose; to Michael Usher for encouragement and advice on writing a book; and to the publishers at Cambridge University Press. Finally, Dave Richardson provided the impetus for the book – without his enthusiasm it would not have been written.

A substantial part of this work is based on research supported by the National Research Foundation of South Africa (UID# 85412), and the DST-NRF Centre of Excellence for Invasion Biology.

Abbreviations

AWRA: Australian Weed Risk Assessment scheme, see Gordon *et al.* (2010) for an overview
CBD: Convention on Biological Diversity, a global agreement addressing all aspects of biological diversity (www.cbd.int)
CWMA: Cooperative Weed Management Area (Box 8.1)
EDRR: Early detection and rapid response
EICAT: Environmental Impact Classification for Alien Taxa, see Hawkins *et al.* (2015)
EPPO: European and Mediterranean Plant Protection Organisation (www.eppo.int)
FAO: Food and Agriculture Organisation of the United Nations (www.fao.org)
IPPC: International Plant Protection Convention (www.ippc.int), overseen by the FAO
ISPMs: International Standards for Phytosanitary Measures, produced by the IPPC
IUCN: International Union for Conservation of Nature (www.iucn.org)
RPPO: regional plant protection organisation

1 · *Introduction*

Campuloclinium macrocephalum (or pom-pom weed) is a South American asteraceous herb that was probably introduced into South Africa in the 1950s as a garden ornamental. It was first recorded as a **naturalised weed** (see the glossary for this and other terms; terms in the glossary are indicated in bold on the first usage in the text) in the 1960s and was still present at low levels well into the 1990s. But over the past 20 years it has spread throughout the biodiverse grassland biome and beyond (see cover image). Every summer the fields between Pretoria and Johannesburg turn pink from the inflorescences of pom-pom weed. The plant creates near-monocultures, reducing biodiversity and the land available for subsistence grazing. Since 2008 there have been concerted efforts to clear populations and limit further spread. While this has, in places, reduced the abundance of some populations, there are no documented examples of even fairly small populations of 1–10 ha having been **extirpated**, and, given the rise in sightings, it is clear that the species has continued to spread in extent (Wilson *et al.* 2013).

The history of **plant invasions** abounds with such examples – the proverbial train-crash in slow motion. Over the last century pine invasions have covered many areas of New Zealand and South Africa, and similar invasions are now developing in South America (Richardson, van Wilgen, & Nunez 2008; Simberloff *et al.* 2010). Leafy spurge and salt cedars continue to spread across North America. The damaging effects of invasions by temperate acacias will likely be replicated in the tropics with a different cohort of invaders (Richardson, Le Roux, & Wilson 2015). The marine alga *Caulerpa taxifolia* was first detected in the Mediterranean in the 1980s, but no action was taken. It is now considered one of the world's worst invaders, known as the 'killer algae'. But in stark contrast to the situation in the Mediterranean, it took only 17 days from the first report of *C. taxifolia* in California until control measures were applied (Anderson 2005). The invasion has subsequently been declared eradicated (Simberloff 2009). In many cases relatively simple (though intensive) actions will limit future widespread consequences.

"We never should have waited this long ... Now the weeds have *completely* taken over."

Figure 1.1. Invasive plant management has been reactive or preventative pre-border rather than proactive post-border. Reproduced with permission from Leigh Rubin and Creators Syndicate, Inc.

Plant invasions often start slowly. Sometimes a few individuals are introduced and so it takes time to build up numbers. Sometimes there are specific limiting factors, so population growth is initially slow (referred to as a lag phase). Often, there is a significant opportunity to mount an effective response. But ironically, the fact that there is time to respond can mean that nothing happens until **eradication** and **containment** are no longer feasible **management goals**, as plant invasions are often difficult and in some cases impossible to eradicate once established. The management of plant invasions is typically either reactive (an invasion is already well established, and the impacts unequivocal before action is taken) or preventative (applying measures that prevent an alien plant from entering a country or ecosystem). There has been much less effort to control invasive and potentially invasive plants after they have established but before they get out of hand (what we term here as **incursion response**), not least because it is a challenge to secure the necessary funding and resources to respond to an incursion before there are large and palpable negative impacts.

From a management and policy perspective there has been a growth in interest in proactive management of plant invasions. More countries are developing **biosecurity** and **plant health** programmes that include detailed strategies on how to manage and reduce the impacts of invasive plants. Global agreements, for example the Convention on Biological Diversity and the International Plant Protection Convention, provide provisions and guidelines on how to address invasive plants, in many cases prior to arriving at a country's borders.

While proactive management is increasingly a focus of invasive plant programmes, it is still difficult to predict which invasions will happen where and when. We simply don't know. But what we do have is space-for-time substitutions. Incursions in one country or area can provide us with insights into what might happen elsewhere. For example, it is believed that the most robust predictor of whether a plant will become invasive is whether it is invasive in another region with a similar climate. For policy makers and land managers, responding to alien plants proactively is still, to a large extent, a matter of attempting to determine the highest-risk species, and then addressing these priority species based upon existing capacity and resources. Given that a number of potentially invasive plants might concurrently be undergoing transitions in status from casual to established, and from established to invasive, which plants should be targeted for coordinated management? This is a question that falls under the discipline of post-border weed risk management, which has been formalised as a protocol in order to foster further development of decision support models for **prioritising** species for management at different jurisdictional levels.

The ability to respond proactively has several challenges. A particular plant species (e.g. pom-pom weed) might never have been recorded as invasive anywhere else in the world and so there is no precedent for predicting impact. In such cases there will likely be low levels of awareness of the species among managers, and little information on how to respond. In other cases it might take decades before population-level processes lead to a widespread invasion; for the net impacts of an invasion to be substantially and demonstrably negative; and for societal views to change regarding the threat posed by a species (van Wilgen &

Richardson 2014). While this creates an opportunity for proactive management, potential deleterious impacts are often discounted or ignored as the costs are temporally and spatially far removed from the site of introduction. Finally, given that plants might be found across several administrative regions, it is not always clear who should champion proactive management efforts. One solution is to conduct a **risk analysis**. Potential problems are anticipated (risk identification), the likelihood and consequences of an invasion are predicted (**risk assessment**), and explicit management and regulatory measures are recommended (**risk management**). However, many authorities and jurisdictions do not have the capacity to conduct a risk analysis, which limits their ability to apply proactive management measures. Fortunately, there is a growing paradigm shift. Regulators, managers, and scientists are beginning to appreciate and act on invasive plant problems before they get out of hand – in many cases before they enter a country or ecosystem, by following the age-old adage, 'an ounce of prevention is worth a pound of cure'.

This book is about the theory and practice of responding to **alien plants** before they become widespread invaders. Would it have been possible to predict that pom-pom weed was going to invade the way it did in South Africa (Chapter 2)? Could it have been detected earlier through better **surveillance** (Chapter 3)? What should have been the national management goal (Chapter 4)? How could progress towards this goal have been measured (Chapter 5)? How could legislation and **regulation** best be used to facilitate pom-pom weed management (Chapter 6)? What **strategy** should have been taken, and what sort of **action plan** was needed (Chapter 7)? And finally, and perhaps most importantly, what organisational structures were needed for this to happen (Chapter 8)?

The general model of the invasion process has, in various forms, been well documented. For the purposes of this book we consider four stages of invasion (**pre-introduction**, **incursion**, **expansion**, and **dominance**), each linked to a specific **management goal** (**prevention**, eradication, containment, and **asset protection/impact reduction**). These specific management goals are often couched only in terms of single species (a species-based approach), but should also consider managing priority areas invaded by one or more species (an area-based approach), or managing the **pathways** that are responsible for the spread of alien species (a pathway-based approach) (Fig. 1.2).

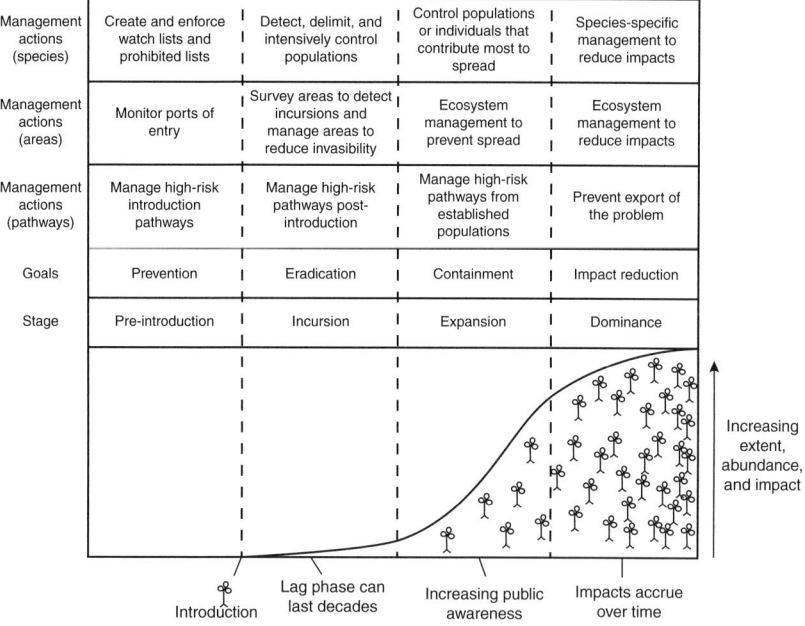

Figure 1.2. A conceptual framework for the management of biological invasions. There is a plethora of similar schemes that one could adopt, but we recommend that any scheme should be careful in separating the invasion stages from the overall management goal and the proposed actions. This framework is adapted from that used as the basis of South Africa's National Strategy for Biological Invasions (Box 7.1).

The focus of this book is therefore primarily on the incursion stage. Importantly, populations that are casual, naturalised, or invasive might all be incursions, but widespread invasions are not. However, whether a population is casual, naturalised, or invasive will affect the most appropriate incursion response (Table 1.1).

1.1 A Brief History of Incursion Response

There are some excellent examples of incursion response programmes that have reacted quickly and efficiently to new invasions well before

Table 1.1. *Recommended response to different categories of alien plants. The categories are as defined by Richardson* et al. *(2000b).*

Category	Definition	Response
Casual	Alien plants that might flourish and even reproduce occasionally in an area, but which do not form self-replacing populations, and which rely on repeated introductions for their persistence.	Monitor (look for evidence of naturalisation) Only consider further action if resources are available
Naturalised	Alien plants that reproduce consistently and sustain populations over many life cycles without direct intervention by humans (or in spite of human intervention); they often recruit offspring freely, usually close to adult plants, and do not necessarily invade natural, semi-natural, or human-made ecosystems.	Monitor (look for evidence of spread) Conduct risk analysis if invasive elsewhere and climate is suitable* Consider management feasibility and start implementation
Invasive	Naturalised plants that produce reproductive offspring, often in very large numbers, at considerable distances from parent plants (approximate scales: >100 m in <50 years for taxa spreading by seeds and other propagules; >6 m in 3 years for taxa spreading by roots, rhizomes, stolons, or creeping stems), and thus have the potential to spread over a considerable area.	Conduct risk analysis Consider management feasibility and start implementation

* Climatic suitability (and other factors) should be considered over the entire jurisdiction potentially affected. Conditions might be marginal where the introduced plant is detected, but much more suitable elsewhere. Non-suitable local conditions should increase the feasibility of management, if an intervention is considered warranted.

there were large negative impacts. There are over 700 documented eradications of vertebrate species from islands (http://diise.islandconservation.org, accessed 13 May 2014), and an increasing number of successful campaigns against plants (though not as systematically documented). One country that arguably has been leading efforts at incursion responses against plants is New Zealand (Box 1.1).

Box 1.1 *Incursion Response in New Zealand (Philip E. Hulme)*

In New Zealand, over half of the entire flora is composed of naturalised alien plants, and this level of invasion has considerable economic and ecological consequences (Hulme 2014). For example, invasive plants threaten one-third of all of New Zealand's nationally threatened plant species and estimates suggest that, without action, invasive plants would degrade the conservation estate, corresponding to a loss of native biodiversity equivalent to NZD 1.3 billion. The costs to the productive sector are easier to quantify and, not surprisingly, much higher, as is the case for the pastoral sector, where the aggregate annual cost of invasive plants has been estimated to be around NZD 1.4 billion. Many of the most problematic plant species in New Zealand became established in the nineteenth century, soon after European settlement, and are now sufficiently widespread that eradication is no longer viewed as an option. Nevertheless, for species that are less widespread, the National Interest Pest Response programme established in 2006 aims to eradicate selected established invasive plants from New Zealand. Species were selected for national response because of their potential to have a significant impact on economic, environmental, social, and cultural values. The final selection of ten alien plant species (Box 1.1 Table 1) was the result of representatives from the Ministry of Primary Industries, Department of Conservation, and local government bodies undertaking a one-off prioritisation exercise that considered the technical, practical, cost–benefit, strategic, and acceptability aspects of each species. In most cases, the response entails intervention to eradicate the species from New Zealand, but in two cases this also includes containment to either the North or South Islands of New Zealand.

The strategy underpinning the National Interest Pest Response programme is to manage existing known sites where species are found, but also respond promptly to any newly discovered populations or new incursions. Eradication programmes involve continued management though the application of physical, chemical, or, in the case of *Hydrilla verticilata*, biological control (using grass carp) that aims to remove the target species until zero density is achieved. This is followed by surveillance at each site for several years, dependent on the likely size of the seed bank or risk of regeneration from rhizome fragments. The goals for both hornwort and Johnson grass have been achieved, and in both cases the programme has shifted to one of surveillance. Other species, such as white bryony and Cape tulip, have

Table 1. *Alien plants listed under New Zealand's National Interest Pest Response programme and the response goal*

Common name	Scientific name	Family	Goal
Salvinia	*Salvinia molesta*	Salviniaceae	Eradication
Water hyacinth	*Eichhornia crassipes*	Pontederiaceae	Eradication
Johnson grass	*Sorghum halepense*	Poaceae	Eradication
Cape tulip	*Moraea flaccida*	Iridaceae	Eradication
Pyp grass	*Ehrharta villosa*	Poaceae	Eradication
Phragmites	*Phragmites australis*	Poaceae	Eradication
Hydrilla	*Hydrilla verticillata*	Hydrocharitaceae	Eradication
White bryony	*Bryonia cretica* subsp. *dioica*	Cucurbitaceae	Eradication
Hornwort	*Ceratophyllum demersum*	Ceratophyllaceae	Eradication and exclusion from the South Island
Manchurian wild rice	*Zizania latifolia*	Poaceae	Eradication of isolated populations and containment of large populations on the North Island

been more challenging; seed longevity and dispersal of these species have hampered eradication. In the case of Manchurian wild rice, several sites targeted for eradication occur within production forests, and require ongoing, close liaison with the owners to ensure the pests are not spread as a result of forest management activities. Water hyacinth and salvinia have had the additional challenge that members of the public have been deliberately spreading these species. However, any new sites are contained to prevent the weeds spreading further and are treated with herbicide. Subsequently, the sites are inaccessible to the public until the complete eradication can be confirmed. It is hoped that for many species the goal will be achieved by 2021.

However, incursion response is not restricted to species listed in the National Interest Pest Response programme. New Zealand maintains a register of 'unwanted organisms' which are understood to be capable of causing harm to any natural and physical resources or human health. Although government has no obligation to act against an unwanted organism simply because it has that status, in certain cases an incursion of an unwanted organism results in an eradication campaign. Once an organism has been detected, an incursion response is initiated to stop or restrict the spread of the organism, identify it, and define its distribution ('delimitation'), followed by an assessment of management options – including control or eradication. The following three examples highlight both the kinds of plant species recently targeted for eradication and the different approaches to managing the problem.

(1) Sea anemone passionflower (*Passiflora actinia*) is a potential threat to New Zealand's environment, with its ability to smother and shade the vegetation it grows on. It has been present at a former horticultural nursery since 1993, the only known population of this species in New Zealand. Initial treatment of the passionflower vines occurred in mid-2012. Over a two-year period, negotiations with the property owners resulted in an agreed cost share arrangement to eradicate the species and replace the shade house roof upon which many of the vines occurred. Eradication was declared in May 2014 and the surveillance operation continued until November that year.

(2) Blackgrass (*Alopecurus myosuroides*) is a serious weed that affects winter crops in Europe, resulting in yield loss, and could have an economic impact on New Zealand agriculture if it became established. Unfortunately, an estimated 2100 blackgrass seeds were spilt from a contaminated consignment of red fescue seed along a 40 km route in the South Island in 2013. An industry–government partnership was established to address this incursion, including media releases on national television. The operation included ten rounds of surveillance, three rounds of mowing, and targeted applications of a selective herbicide to high-risk sites along the entire 40 km route. It is proposed operations and surveillance will continue for three years but may be reviewed earlier if the risk is sufficiently reduced. The liable company admitted fault and has been active and supportive in the response process, fully funding the operational activity and providing the expertise to carry out the herbicide application. A check system for transport operators has been implemented to help ensure every load is secured appropriately before transport to prevent further spills.

(3) Sea spurge (*Euphorbia paralias*) is a European dune shrub that was introduced to Australia in ballast water and now forms dense stands in foredune and backdune habitats, threatening native biodiversity. The species is now widespread across the south coast, from Perth to Tasmania, and its ability to spread on ocean currents means

Figure 1. A publicity fact sheet produced by the New Zealand government to improve passive detection and communicate management activities.

there is a risk it will reach New Zealand, where it could become a serious coastal environmental weed. Sea spurge was first detected in New Zealand in February 2012 and over the following two years an eradication programme was successfully completed. The detection site remains the only known site of sea spurge in New Zealand, but clearly the risk of further incursions remains (Box 1.1 Fig. 1).

These three examples reveal how incursions are driven by different introduction pathways, the importance of having a register of unwanted species and the key role the public plays in detecting new incursions, as well as involvement in eradication through partnerships and cost-sharing agreements.

A key feature of a successful incursion response is the ability to respond quickly – a good example is the response to *Caulerpa taxifolia* in California, where treatments began 17 days after detection (Anderson 2005). Having capacity available that can respond to emergencies is vital, and is an important feature of the planned response to *C. taxifolia* should it be detected again in California. This approach is often termed **early detection and rapid response (EDRR)**. EDRR should be incorporated into management at all stages of the invasion process, although it is often used synonymously with the control of incursions. However, as for management at other stages of the invasion process, EDRR should not be the sole approach for controlling incursions. Control measures often need to be sustained over long periods and the strategy used adapted to what is actually happening on the ground (Simberloff 2009). Eradication often takes years or decades of persistent, dedicated, and well-organised effort, requiring continual surveillance and a sustained response. A notable example is that of *Striga asiatica* in North America. The extent of this parasitic weed of agricultural systems has been reduced significantly over 50 years of dedicated management efforts, such that the invader is now contained in a small area in North Carolina as part of an ongoing campaign aimed at achieving continent-wide eradication (Tasker & Westwood 2012). Another parasitic weed, *Orobanche ramosa*, was also effectively contained in South Australia, although in this case as part of an eradication programme in South Australia that was ultimately unsuccessful (Panetta 2012). Effective incursion response will sometimes be the hare and other times the tortoise.

The largest and most valuable eradication of a **biological invasion** to date was the global effort to eradicate smallpox. One of the key lessons learned was that invasions are specific to a particular point in space and time, so management needs to be context-specific. Control had to change from country to country depending on the socio-political and economic context. While success was ultimately reliant on political will, such efforts required overall dedicated coordination: 'The significant improvement in efficiency and supervision that can often be realized in special programmes may well offset the additional costs they sometimes entail' (Fenner *et al.* 1988, pp. 1351–1352).

Can the lessons from these examples be applied more widely? Given the growth in theory and practice, there is cause for optimism.

Invasion science has grown as a discipline over the last 50 years, from a largely biological domain to a multi-disciplinary science. Several of the top applied environmental and ecological journals (e.g. *Ecological Applications*, *Journal of Applied Ecology*, and *Diversity and Distributions*) have a significant number of papers on invasions; there is a growing number of journals dedicated specifically to invasions (e.g. *Biological Invasions*, *NeoBiota*, *Management of Biological Invasions*, and *BioInvasions Records*); and several journals specific to the management of plant invasions and weeds (e.g. *Aquatic Plant Science and Management*, *Invasive Plant Science and Management*, *Weed Biology*, and *Weed Science*).

The theory behind eradication and population **delimitation** has also developed significantly over the past few decades, providing important management insights. For example, there have been notable advances in models used to address the question of how to allocate management resources across the landscape of an invasion. In a seminal paper, Moody and Mack (1988) produced a simple model to highlight that under many conditions the best way to reduce the rate at which an invasion occupies more area is to control small satellite populations. In one of several further developments of the concept, Taylor and Hastings (2004) found that if the growth rate of small populations is relatively low (i.e. there was an Allee effect, where individual fitness declined at low population sizes or densities), then the optimal strategy depends on the resources available. If management resources are limited it is best to target small satellite populations and reduce spread. However, if sufficient resources are available, eradication would be achieved more quickly if the core population was also targeted. This approach was applied in practice to an invasion of *Spartina alterniflora*, where isolated patches are known to produce fewer propagules per unit than high-density coalesced meadows. Additional considerations will arise, e.g. if the minimum generation time is long and only old plants produce large numbers of seeds, then it might be best to tackle the core population first. But the general approach to the problem still holds. Other approaches have also provided useful insights. For example, Chadès *et al.* (2011), using network analyses, presented simple and robust rules of thumb for the order in which to manage subpopulations connected in different general patterns (e.g. directed lines or connected clusters).

While many such population and economic models are theoretical, and are largely untested, the rise in the number of on-ground case studies means that it is increasingly possible to evaluate the success of actual eradication campaigns. For instance, Rejmánek and Pitcairn (2002) showed that in California major initiatives to extirpate populations over 1000 ha almost always failed, while efforts to extirpate populations of less than 1 ha tended to succeed, often with little cost. In a review of 173 eradication campaigns against a variety of taxa, Pluess *et al.* (2012b) similarly found the spatial extent of an incursion to be a critical factor in determining the likely success of a campaign. And in a recent review of efforts to control plant incursions in Victoria, Dodd *et al.* (2015) found the detectability period, propagule longevity, and time to reproductive maturity were as, if not more, important than population size in determining the success and time to extirpation. Further work has categorised species in terms of how difficult they would be to eradicate at the same invasion stage (Panetta 2015), specifically using time to reproductive maturity, propagule longevity, and dispersal syndromes as the key determinants (Fig. 4.6).

A major limitation recognised by such analyses is that the variables used tend to be limited to those that are easily accessible and comparable (e.g. species traits), rather than those that might well be more important (e.g. organisational structure and management context). We define **eradication** and **containment feasibility** in terms of features of the individual species, the environment where the incursion is happening and how a species has been introduced (i.e. incorporating aspects of invasiveness, invasibility, and introduction dynamics). But as pointed out by Paul Gobster (2011), 'sensitivity to the social dimensions of the issue may be as important as understanding the ecological and technical aspects of management'.

1.2 Invasion Debt

The central focus of this book is how to develop coordinated methods to reduce future problems by acting now. The concept of the size of

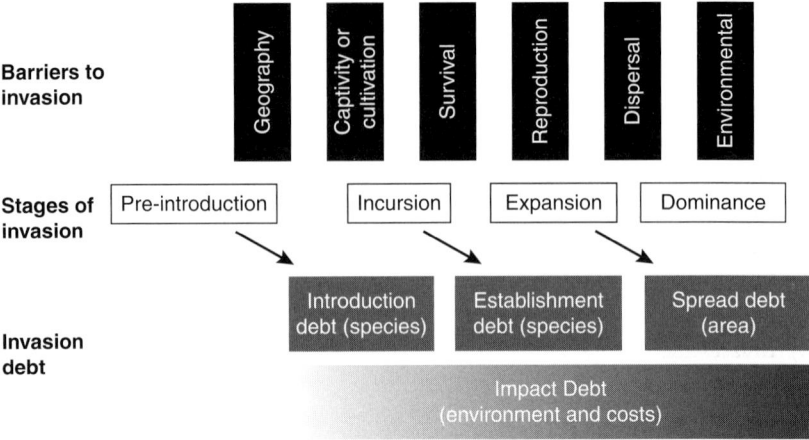

Figure 1.3. Invasion debt. Current impacts of invasions are due to processes that happened decades or centuries ago. Similarly, future impacts will arise from current processes, i.e. we are facing a significant 'invasion debt'. By quantifying potential threats, the value of proactive management becomes apparent and the relative values of prevention, eradication, containment, and impact reduction can be assessed. The figure is adapted from a conceptualisation of the invasion process as a series of stages whereby barriers to invasion are overcome (Richardson *et al.* 2000b; Blackburn *et al.* 2011), with a debt associated with each stage. Introduction debt and establishment debt are measured in terms of numbers of species; spread debt is measured in terms of area; and impact debt can be measured in terms of economic values and estimates of biodiversity loss. Redrawn from (Rouget *et al.* 2016) with permission.

the future problem due to biological invasions has been formalised as 'invasion debt' (Rouget *et al.* 2016). This is composed of several key components (Fig. 1.3).

(1) Species will keep on being introduced (the **introduction debt**). The rate of introduction of species by human-mediated dispersal is thousands of times that of historical rates of natural dispersal (Ricciardi 2007). Most of the introduced plants will simply add to the non-native biodiversity of a region, some will have undesirable properties without becoming invasive (e.g. many eucalypts in South Africa and California are not invasive but can tap into and exhaust groundwater), while a few others will progress to become invasive and have impacts (in many cases both negative and positive).

(2) Species already introduced will naturalise and become invasive in the future (**species-based invasion debt**). Using records collected by

the botanist Harold Lyon on Hawai'i, Daehler (2009) found the average time between introduction and naturalisation to be 14 years for woody plants and five years for herbaceous plants. Harold Lyon set out to monitor such patterns, so it can be assumed there was little delay between actual naturalisation and the first record of naturalisation. But mean estimates from South Australia of time from introduction to the first record of naturalisation are in the order of decades for herbaceous perennials to over a century for woody perennials (Caley, Groves, & Barker 2008). There are various reasons for these temporal delays (Chapter 2).

(3) Invasive species will increase in range (**area-based invasion debt**). The current distribution of most invasive species is still strongly influenced by the number of sites of introduction (Donaldson *et al.* 2014). Estimates from modelling current and potential distributions suggest that it will take centuries for most invasive species to come close to reaching their full broad-scale potential distribution (Wilson *et al.* 2007; Williamson *et al.* 2009).

(4) The impacts from invasive species will increase over time (**impact-based invasion debt**). The longer a given area is invaded, the more likely it is to experience biotic and abiotic changes as a result of the invasion. Initially there might be altered species composition, but vegetation and seed banks are intact and the community is still very close to a native ecosystem state. At a certain level of invasion a biotic threshold is crossed, resulting in altered species composition and structure. Finally, if abiotic changes (e.g. altered water and nutrient availability) accrue over time, there can be positive feedbacks leading to changes which are in practical terms irreversible, in effect creating a novel ecosystem (Hobbs *et al.* 2006; Gaertner, Holmes, & Richardson 2012).

When measured over ecologically and management-relevant time scales (e.g. 20 years), invasion debt can decline as well as increase. Management interventions can block dispersal, effectively reducing rates of introduction (e.g. ballast water management efforts on the Great Lakes of North America have largely prevented new introductions via ballast water (Bailey *et al.* 2011)); eradications will reduce the number of alien species in a region; efforts at containment will limit the area invaded; and control, mitigation, and utilisation can tip the impacts caused from negative to positive. Predicting which species introduced in the future will become invasive and have an impact, is a major area of invasion science. At a direct, applied level such analyses inform management pre- and at-border and are the basis for **watch lists** (to prevent accidental or deliberate illegal introductions) and **permitted** and **prohibited lists** (to restrict legal introductions to only those that pose an acceptable risk).

We suspect that in many regions we are only witnessing the start of the invasion process. Most of the future invasive plant species in a region will likely come from the pool of those already introduced, but management efforts have focused on preventing new introductions and on widespread invaders already causing large impacts. There should be more focus on incursion response.

1.3 Main Steps and Take-Home Messages

We consider there to be three main steps in incursion response planning (Fig. 1.4): (1) predict, prevent, and prepare; (2) determine management options and take action; and (3) evaluate performance and revisit goals if required. These aspects are covered in the first part of the book, where we focus on identifying the problem and determining and evaluating the appropriate response (Chapters 2–5). In the second part of the book we concentrate on the enabling environment (i.e. legislation, strategies, action plans, and processes necessary for implementation) (Chapters 6–8). We conclude with some final thoughts and considerations of where the field might (or should) be going and summarise the key findings (Chapter 9).

As we see it, the main take-home messages are:

- Management of alien plant incursions can and should be proactive.
- Specific data are required in order to make informed management decisions. Such data are not always available, so protocols for data collection need to be implemented.
- Management decisions need to be flexible and updated, depending on the documented progress.
- An iterative process tightly linking research and management is often an essential component of incursion response.
- Successful alien plant incursion response depends both on collaboration and a dedicated champion.

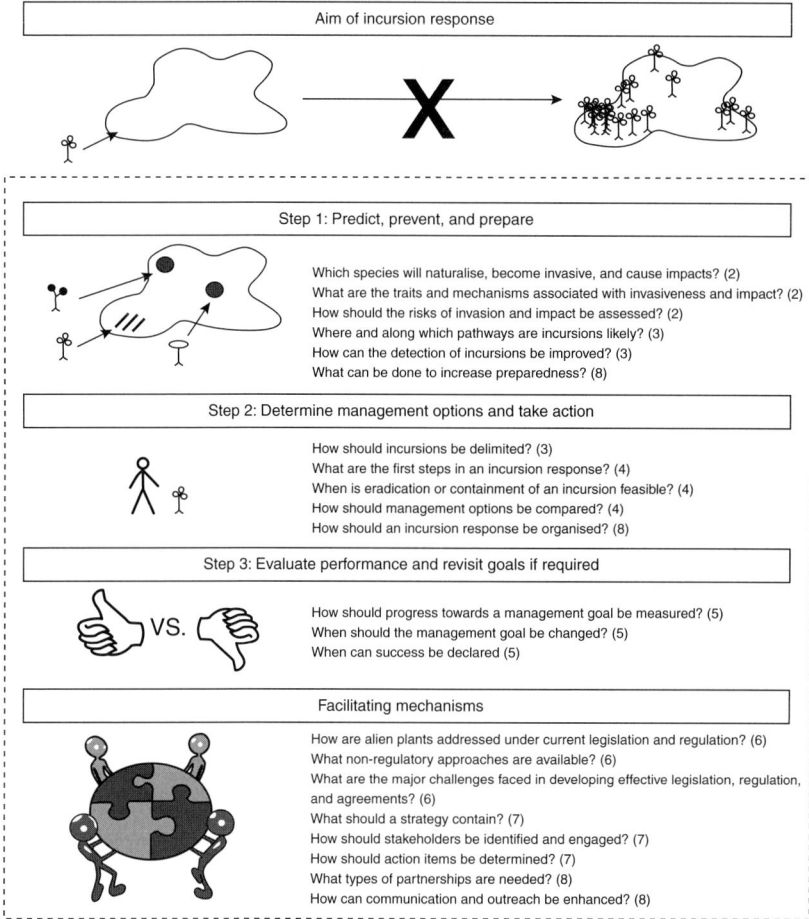

Figure 1.4. The basic process for an incursion response. The aim of incursion response is to prevent an alien species from becoming a widespread invader by taking proactive measures. In this book we consider three main steps to achieve this in the context of a series of facilitating mechanisms. The key questions covered at each step are shown, with the relevant chapter indicated in parentheses. In Chapter 9 we revisit this and discuss recommendations at each step (Fig. 9.1).

1.4 Definitions

We include a glossary of key terms, endeavouring to use the most commonly used definitions throughout, though in some cases there is disagreement (even among ourselves). Therefore in the glossary we highlight instances where definitions vary. For instance, in this book we reserve the word *eradication* for the removal of all individuals and propagules of a species from a defined area to which the likelihood of recolonisation can, for all practical purposes, be ignored (Myers, Savoie, & van Randen 1998). This is largely the same as the definition under the International Plant Protection Convention (ISPM No. 5) – the 'application of phytosanitary measures to eliminate a pest from an area' – but the term *eradication* is also often used as a synonym for applying intensive control efforts; it literally means 'pulling up by the root', from the Latin *eradico*. We prefer to stick with a strict definition of eradication as (in the sense used here) it is qualitatively different from other management goals.

On this note, there is an ongoing debate about how to define biological invasions. This debate is largely moot as there are three distinct concepts that are important – biological invasions as a biogeographical phenomenon, as an expression of ecological dominance, and as interpreted through human values. For often seemingly good reasons these definitions are frequently conflated, and this has led to confusion at the interface between management, science, and policy. Throughout the book we have tried to be consistent: alien, naturalised, and invasive species refer solely to the biogeographical definition; when describing ecological dominance we try to be explicit in terms of measures of abundance or spread relative to other components of the ecological community; the terms *weeds* and *pests* are used where a specific value judgement has been made (by individuals or society as a whole). To put it another way, all invasive plants are alien, but they aren't all 'bad'; weeds, whether native or alien, are.

2 · *Prediction (Pre- and Post-Border)*

Key questions addressed:

- Which species will naturalise, become invasive, and cause impacts?
- What are the traits and mechanisms associated with invasiveness and impact?
- How should the risks of invasion and impact be assessed?

One of the authors of this book lives and works in Stellenbosch, a town close to Cape Town on the south-western tip of Africa. The landscape is covered in unique types of vegetation – renosterveld and fynbos. Or at least it was. Now Mediterranean pines are scattered over the mountains, Australian wattles cover many of the lowlands and waterways, and the streets of Stellenbosch are lined with European oaks. But these are not the most visibly striking introduced plant species in and around Stellenbosch. The reason Stellenbosch became wealthy (and that oaks were introduced) was wine production. The surrounding hills are covered in vineyards. But grapes are clearly not invasive, despite being bird- and human-dispersed, and planted in massive quantities. Of course this is a trite example: cultivated grapes tend to be sterile and they are highly domesticated, often requiring substantial agricultural inputs. The pines and wattles are widespread as they were planted in massive numbers, they produce copious amounts of seed, they are well adapted to the natural fire regimes in the area, and the fynbos has no fire-adapted tree species. It is easy to generate convincing post-hoc explanations, but it is much harder to predict which species will be the next invasive weed.

Understanding which introduced species will become invasive and cause impacts is one of the core goals of invasion science. But despite years of research, there are few general robust predictors beyond the observations that those species that are likely to become invasive are those that are invasive elsewhere in the world, that come from areas with a similar climate, and that are likely to be introduced (Hulme 2012). This is unsurprising, as biological invasions are complex phenomena that can

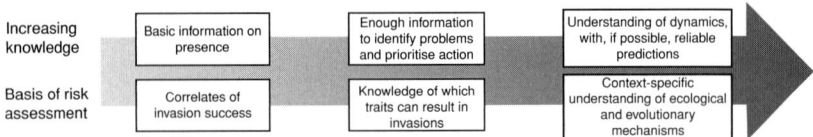

Figure 2.1. As our understanding of a situation improves, so the decisions made can improve. Based on Wilson *et al.* (2014) and Novoa *et al.* (2015a).

have multiple causes. More sophisticated management decisions can be made if we understand the biotic and abiotic mechanisms operating for a particular system (i.e. there can be substantial value in working on a case-by-case basis), but in many cases decisions have to be made with little information available (Fig. 2.1).

2.1 Correlates and Traits Associated with Invasiveness

The first part of the problem is to predict which species might become invasive. Here we discuss the key theory underpinning this, and, in particular, note that the factors used to predict invasion success can be split into three broad categories: biogeographic, human influence, and biological.

One of the most common biogeographical correlates of invasiveness is native range size. Species with large native ranges tend to be invasive more often (Williamson & Fitter 1996). The mechanism for this is unclear: species with larger ranges might be encountered more often and thus be more likely to be transported; they might occupy a wider range of environmental conditions; or they might inherently have higher population growth rates at low densities and more effective dispersal. Such biogeographic analyses have been expanded to look at how abundance–occupancy patterns in both the native and introduced ranges can be used to predict population-level dynamics (Wilson *et al.* 2004; Veldtman, Chown, & McGeoch 2010; Hui *et al.* 2011). In comparison to coarse measures of total range size, there is likely to be a closer mechanistic link between the abundance–occupancy patterns and population dynamics, although much of this can be due to introduction dynamics (Donaldson, Richardson, & Wilson 2014).

Invasions, by definition, are the result of human influence. As we will discuss in Chapter 3, both the dispersal pathways into and around a country, and the resulting location of invasions are contingent upon the introduction dynamics (i.e. human influence) (Wilson *et al.* 2009). One of the clearest consequences of this is the often overriding importance of

propagule pressure. The more a species is introduced, the more likely it is to become invasive: the more widely a species is planted, the larger its invasive range. Often it can be difficult to know the exact number of propagules released, but several correlates are available, including numbers of seeds imported and the frequency with which species appear in various catalogues and databases (Pemberton & Liu 2009). The importance of propagule pressure is such that it has been proposed as a null model for explaining differences in the probability and extent of invasions (Colautti, Grigorovich, & MacIsaac 2006). There is no need to evoke biological or biogeographic traits if one species has simply had many more opportunities to invade than another.

Analyses of invasions have, however, tended to focus on biological factors. Baker (1965) produced a list of traits he thought defined an ideal weed, including rapid seedling growth rates, short juvenile period, self-compatibility, adaptation for long- and short-distance dispersal (see van Kleunen, Dawson, and Maurel 2015 for a recent review). Thirty years later, Rejmánek and Richardson (1996) examined ten life-history traits for predicting the invasiveness of pines (genus *Pinus*). They took a multivariate approach and found that invasive and non-invasive species could be separated based on an interaction between juvenile period, seed mass, and the interval between large seed crops. The relationship proposed was in essence the result of variation from r-strategy species (likely to become invasive) to K-strategy species (likely not to be invasive). A heuristic approach like this is valuable, and can help develop hypotheses, but the link to r–K strategy could be based in part on the rate at which invasions are observed. A short juvenile period means species will probably become invasive sooner and can make invasions more difficult to manage (Chapter 4). However, a longer juvenile period does not limit an invasion per se, and the juvenile period need not be linked either to eventual rates of spread (likely more a function of dispersal distances), or to impact.

One of the most widely studied biological factors is that of phylogenetic relatedness to other invasive taxa. The traits that ultimately result in invasions are sometimes phylogenetically conserved, creating taxonomic correlations to invasiveness. Certain families have a high proportion of invasive species (including the species-rich Asteraceae, Fabaceae, and Poaceae), while other families contain few invasive species (e.g. Bromeliaceae and Orchidaceae) (Daehler 1998; Pyšek 1998). This does not, however, provide sophisticated information for regulation, since, for example, the option of banning all grasses is not compatible with global agriculture. Analyses need to be at a much finer phylogenetic scale for decision makers. This has led to attempts to identify lineages with a very high percentage of invasive species.

From the standpoint of the recipient environment, one might also expect that alien plant taxa are more likely to succeed where similar species are already present, since the former would be more likely to share the same success-assuring traits. However, in testing this idea with European plants introduced to North America, Charles Darwin found the opposite result. He posited that if a species shares evolutionary history with taxa in an introduced area, then there might be greater resistance to invasion – an idea that is now termed Darwin's naturalisation hypothesis, with the mechanism now known as biotic resistance. There have been several empirical tests of the phylogenetic pattern of invasions since (e.g. Procheş *et al.* 2008). The results have varied, but are consistent with the idea that the phylogenetic patterns in biological invasions change depending on the spatial and taxonomic scale, since different mechanistic processes (e.g. competition versus phylogenetic conservatism in climatic tolerances) operate at different scales.

Another main correlate used to explain invasive success has been invasiveness elsewhere. If a species has become invasive somewhere in the world, then it is much more likely that it will become invasive if introduced elsewhere. This assumption is the basis of most preventative measures, and is often used to prioritise management. However, one of the most fascinating and potentially productive areas of research in invasion science is comparing the performance of introduced species in different regions. This can provide insights into the drivers of invasions (e.g. native pollinators facilitating invasions in one region but the lack of pollinators preventing naturalisation in another (Richardson *et al.* 2000a)).

Invasiveness elsewhere and the degree of invasiveness in congeners were combined in a recent analysis by Diez and co-workers (2012), where the probability of plant taxa naturalising in New Zealand was predicted based on success in Australia. They found that grass species from genera such as *Lolium* and *Phalaris* would be expected to naturalise, whereas there is a much lower prior expectation that *Hordeum* and *Bambusa* species would naturalise. Explicit methods are needed to incorporate different types of evidence in risk assessments, and in particular to incorporate estimates of the importance and reliability of the different sources. The Bayesian methods used in their study will likely become increasingly common in risk assessments.

The search for correlates of invasive success has not always been fruitful, however. For example, invasive Australian acacias do not form a

2.1 Correlates/Traits Associated with Invasiveness · 23

monophyletic group, and are not strongly clustered phylogenetically (Miller *et al.* 2011). The current major invaders tend to be native in particular Australian agro-climatic classes (Richardson *et al.* 2011), though this is arguably a function of historical biases in which species were used and distributed around the world for forestry (temperate species in particular). Some other traits are over-represented, including biogeographic signals of the native range (Hui *et al.* 2011) and extent of human usage (Castro-Díez *et al.* 2011; Wilson *et al.* 2011), but despite substantial research effort, there is no clear separation between invasive and non-invasive taxa. This has led to the suggestion that all Australian acacias will become invasive species with substantial impacts if they are planted in large enough numbers for long enough, i.e. invasions are based on propagule pressure (Wilson *et al.* 2011). Without any evidence to the contrary this should form the basis of how invasion risk in this group is managed (Richardson, Le Roux, & Wilson 2015).

A major concern with basing management decisions on correlates of invasiveness is that there is likely to be a significant number of false positives – species expected to become invasive based on similarities to known invaders, but which do not become invasive. In such cases imposing a ban seems preposterous to the people who grow and benefit from the plants in question. For example, while some cacti species are widespread invaders, many of the globose cacti species popular in horticulture do not spread vegetatively and none is known to be invasive. There is a variety of traits that separate invasive from non-invasive cacti, but decisions should ideally be made on the basis of traits that are likely to be directly linked to invasiveness, e.g. cacti that disperse via cladodes (Novoa *et al.* 2015a, 2015b).

There have been substantial conceptual advances in the tests of invasiveness. For example, van Kleunen *et al.* (2010) set out a scheme for assessing what is being compared and how to interpret the results. The traits of invasive species can be compared to native species or to non-invasive species; traits of non-invasive species can be compared to native species; comparisons can be made between the introduced and native ranges of the same species; and so on. The take-home message is that the different comparisons need to be interpreted in different ways, and not all of the results can or should directly influence management decisions. One of the main lessons from such analyses has been that different factors can be important at the different stages of the invasive continuum (Fig. 2.2; Dawson, Burslem, & Hulme 2009; Moodley *et al.* 2013). In particular, van Kleunen, Dawson, and Maurel (2015) argue that in order to gain insights

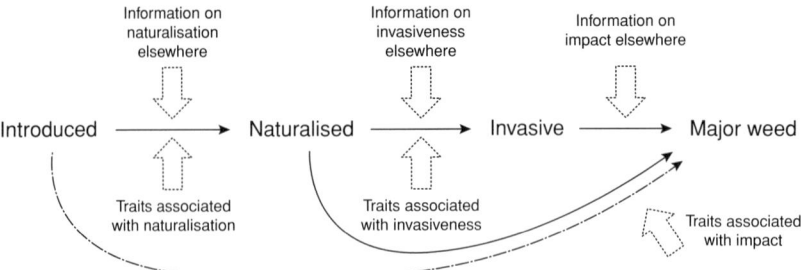

Figure 2.2. Different types of information are needed to make predictions about the likelihood of transition between different stages of the invasion process. Relatively few traits will apply solely to one transition, and some traits can influence the likelihood of all transitions (though potentially in different directions (Moodley et al. 2013)). The transition to major weed differs qualitatively from the others in that it is defined in terms of impacts (environmental, economic, and social) rather than in terms of biological and ecological processes, and, moreover, plants need not be invasive to have significant impacts.

from analyses of traits, such analyses must consider the underlying hierarchical structure of the invasion process.

Research over the past 10–15 years has investigated the key transitions between casual status and naturalisation and between naturalisation and invasion, with the aim of gaining explanatory power, if not a predictive capacity. Traits that are advantageous at an early stage might be neutral or even detrimental at a more advanced stage (Kolar & Lodge 2001; Pyšek & Richardson 2007). For example, analyses of the contaminants of seed shipments have shown there is a selection for weeds with similar seed characteristics to the intended crop (Shimono & Konuma 2008). While this selection for a particular seed size will increase the likelihood of introduction, it will not necessarily facilitate establishment or spread, which might be more influenced by the ability to attach to farm machinery or by an ability to be spread internally via the transport of livestock around a country.

One area where future research is likely to be productive is how the traits associated with invasiveness differ between habitat types (van Kleunen, Weber, & Fischer 2010). It appears highly likely that habitat-specific invasiveness traits do exist. For example, traits that allow rapid use of available resources (as would be important in disturbed habitats) are probably different from those that facilitate invasion in late-successional communities, where resource availability might be considerably lower or highly episodic in nature (Rejmánek 2011).

Another major issue with analyses of invasive traits is that biogeographic, human influence, and biological factors are often confounded. For example, forestry species that are planted in large numbers (human influence) are those that have been selected based on high growth rates (biological) and that are climatically suited to a given region (biogeographic). Similarly, a study in Florida looking at historical records of horticultural catalogues found that plants that became invasive had been marketed for three times as long as plants that did not naturalise (Pemberton & Liu 2009), the likely explanation being that plants that grew well sold more, were planted more, and were therefore more likely to become invasive due to both inherent invasiveness and greater propagule pressure. Invasion success is in large part determined by opportunities for dispersal and the number of propagules involved, which in turn are often under the dominant influence of humans.

So while there has been progress in general understanding of how specific traits relate to invasiveness, there are still few trait-based rules that are useful to those tasked with making decisions relating to the introduction and management of individual species.

2.2 Correlates and Traits Associated with Impact

The second aim of predictive analyses is to work out whether a new invasion will have substantial undesirable consequences. The degree to which traits that confer invasiveness might also confer an ability to generate significant impacts remains an open question (Box 2.1). But at present there is little evidence for a close correlation. For example, Hulme (2012) found no relationship between how widespread invasive plants in the UK are and their estimated economic costs. Similarly, when Ricciardi and Cohen (2007) compared invasiveness (in terms of both rate of establishment and rate of spread) with impact on native biodiversity (ranked categorically based upon documented evidence of a species' effects on the populations of native species), they found that there was no significant correlation for plants, vertebrates, or invertebrates, nor was there a general relationship between ranked level of impact and establishment success rate. The authors suggested that different attributes might determine the invasiveness and impact of a species.

A global assessment of the ecological impacts of invasive plants suggested that a few species traits, namely life form, plant stature, and pollination syndrome could be used to predict species impact across a variety of habitats and geographical regions (Pyšek et al. 2012). Using data-mining methods, in conjunction with classification trees, these authors showed that the most significant impacts were on survival of resident biota, animal activities, community productivity and cover, mineral and nutrient content in plant tissue, and fire frequency. Other outcomes (mostly related to abundance, species richness and diversity, and fecundity of resident biota) were more likely to be significantly impacted if the invasive species was an annual grass. Where other life forms were involved, impact was more likely significant where the invasive species was taller than 4.8 m (i.e. trees and climbers) and the invasion took place in Mediterranean or tropical regions. Resident plant communities on continents were most likely to experience significant reductions in species richness if the invading plant was wind pollinated; if other means of pollination were involved, significant reductions were observed only where the invading plant was taller than 2.8 m.

This study demonstrated significant associations between a relatively small number of plant species traits and ecological impacts over a range of habitats and regions, but it does not negate the general finding that impact, as for invasion, is strongly context-dependent. Recognising the importance of context, a study investigating grass species naturalised in tropical and sub-tropical Australia (van Klinken, Panetta, & Coutts 2013) assessed the impact of introduced grasses against criteria that differed according to the individual sector affected: environmental, agricultural (cropping and horticulture), or pastoral. Fourteen per cent (21/155) of the naturalised species were classified as high-impact, including four species that affected more than one sector. High-impact species were more likely to have exhibited higher spread rates (in terms of regions invaded per decade) and to be semi-aquatic. However, the association of rapid spread with high impact on a local scale was not considered to be explanatory, since spread rate was largely human-mediated: these plants were spread actively by humans over very large areas. The only significant predictor for high-impact weeds of the pastoral sector was life-history (all seven high-impact species were perennial), but the small number of high-impact species limited the scope for further analysis. Other factors, such as alterations of the grass-fire cycle in natural ecosystems and the development of herbicide resistance in agricultural systems, were also expected to be important predictors of impact in individual sectors.

Van Klinken, Panetta, and Coutts (2013) noted the sensitivity of the categorisation process to the criteria that had been employed. For example, the criteria for environmental weeds were relevant to the context of impact, but not to its extent. Their work therefore identified high impact in this sector, but within very restricted settings relative to the size of the regions considered. Since multiple sectors are impacted, it is also important to take into account the potentially competing interests of stakeholders (i.e. various ecological, economic, or social interests), a topic explicitly addressed by Kumschick *et al.* (2012). These authors make the point that the process of using objective, scientific methods to capture the changes associated with a plant invasion must be separated from the comparatively subjective, societal evaluation of impact, which is based on values. We will deal with this topic further in Chapter 6.

The lack of a strong link between invasiveness and impact again implies that recommendations based on invasive traits have limited value to management, and it is therefore important to explicitly consider traits of impact. Obviously, impact is of paramount interest from a management perspective. Unless an introduced plant is capable of generating significant negative impacts under current or likely future scenarios, its ability to establish and spread is largely immaterial. Many plants that are restricted to highly disturbed areas (e.g. transport corridors and industrial sites) might spread rapidly owing to a close association with human activity, but might not have much impact upon the habitats that they invade. For example, some widespread North American forbs such as *Senecio vulgaris*, *Stellaria media*, and *Erodium* spp. have spread widely but are not listed on noxious or wildland weed lists (Drenovsky *et al.* 2012). On the other hand, species might spread slowly and/or episodically owing to ecological or life-history features (e.g. trees that have a very prolonged juvenile period, or water-dispersed plants of semi-arid or arid ecosystems that spread only after rare, high-rainfall events), yet cause serious impacts in the long run. In fact, some species do not spread at all, but still have substantial adverse impacts. For example, many eucalyptus species have not been observed to invade, but, by virtue of their extensive root systems, individual planted trees can have substantial impacts on local hydrology, and so it would be important to consider an incursion response. Predicting which species will become serious pests (as opposed to being invasive per se) is exceedingly difficult and perhaps for this reason has attracted relatively little research attention to date (Parker *et al.* 1999; Pyšek *et al.* 2012; van Klinken, Panetta, & Coutts 2013).

Impact has been defined as a function of range, abundance, and the per-capita or per-biomass effect of the invader (Parker *et al.* 1999). While this conceptual framing of the problem is useful, the scheme has not

Table 2.1. *Broad categories of negative impacts caused by invasive plants*

Impact category		
Environmental	Economic	Social
Biodiversity losses	Decreased agricultural yield	Reduced amenity
Reduced ecosystem services	Cost of control measures	Reduced recreational opportunities
Effects on fire frequency and severity	Marketing limitations on contaminated products	Effects on human health

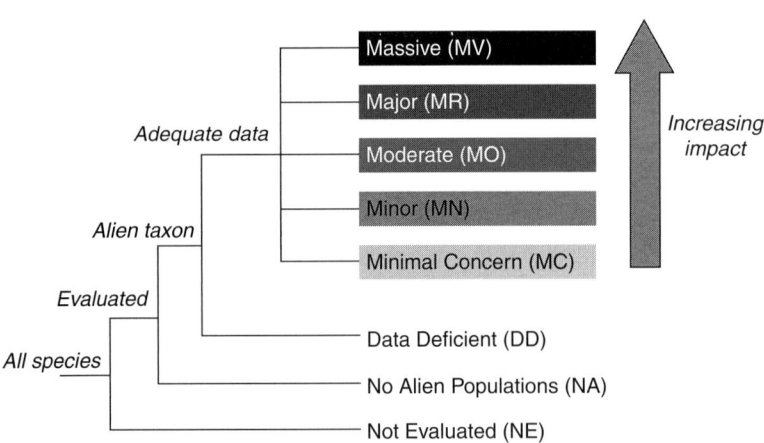

Figure 2.3. The IUCN Environmental Impact Classification for Alien Taxa (EICAT) Scheme. Species are assessed based on documented impacts anywhere in the world, and only given a rating after a set process is conducted (cf. the IUCN Red List Scheme). Importantly, the scheme as it is currently set out is descriptive, i.e. it does not aim to predict impact. Redrawn from Hawkins *et al.* (2015) with permission.

really been implemented in practice. One reason for this is that some types of impact, such as effects upon crop yields, are relatively easy to quantify, but others are multifarious and difficult to measure (Hulme 2012). Not only can plants have significant impacts upon production values, they can also have social impacts (e.g. on human health, recreation and amenity) and environmental impacts (e.g. on biodiversity, ecosystem processes, and ecological services) (Table 2.1). Moreover, it is common for a single species to cause impacts in each of these different categories. For example, *Parthenium hysterophorus* is a significant annual weed of cropping and grazing systems, but also causes human health problems (allergic contact dermatitis, rhinitis, and bronchitis) and has been associated with reductions in grassland biodiversity in many parts of the world (Navie *et al.* 1998).

Various schemes have been developed to rectify this. For example, a framework proposed by Blackburn *et al.* (2014) builds on a general scheme that considers 12 impact classes for biological invasions in terms of biodiversity (of which *competition*; *hybridisation*; *transmission of diseases to native species*; *parasitism*; *poisoning/toxicity*; *bio-fouling*; *chemical, physical, or structural impact on ecosystem*; and *interaction with other alien species* are relevant for plant invasions). Notably, several of these impacts do not require an introduced species to naturalise, let alone become invasive. This impact classification scheme is similar in format to the IUCN conservation Red List, and is in the process of being implemented as the IUCN Environmental Impact Classification for Alien Taxa (EICAT) Scheme (Fig. 2.3; Hawkins *et al.* 2015). The next step is to develop similar metrics for socio-economic impacts. However, much work still needs to be done to resolve the issues of comparing different populations, ensuring that assessments are mathematically sound, developing predictions, and ensuring that the context-specific nature of impacts is appreciated.

Box 2.1 *Plant Traits Associated with Impact on Native Plant Species Richness (Montserrat Vilà, Rudolf P. Rohr, José L. Espinar, Philip E. Hulme, Jan Pergl, J. Jacobus Le Roux, Urs Schaffner, & Petr Pyšek)*

There has been a considerable amount of research on the particular species traits that might determine why an introduced plant species can establish and become invasive. This information is of great value as it can be used as an important component of risk assessment to screen lists of species for introduction (e.g. for gardening, reforestation, biofuel) to identify those that have the potential to become invasive. The general pattern is that invasive plant species are larger and have higher relative growth and physiological rates than non-invasive plants (van Kleunen, Weber, & Fischer 2010). Are these also plant traits that confer greater ecological impacts on the invaded ecosystem? Not necessarily. It is already well accepted that plant success at different invasion stages from introduction to spread are driven by different factors. Different plant species traits play a significant role in each stage, together with characteristics of the ecosystem and the history of introduction. As the success of a non-native species to invade and the extent of invasion are not linked to the damage the invader can cause, traits associated with the success of invasion do not need to be associated with traits conferring impact. Moreover, the impact of many well-known successful invaders has yet to be investigated in depth (Vilà et al. 2011).

Research on plant impacts has mostly focused on assessing the type and magnitude of impacts of non-native species on native plant populations, on plant community structure, or on a handful of ecosystem processes, such as nutrient cycling (Hulme et al. 2013). However, in recent years the first attempts have been made to compile and analyse these studies to provide more generic insights into which plant attributes lead to particular impacts (e.g. Pyšek et al. 2012). More recently, we have conducted a meta-analysis based on 155 studies that looked at the effect of non-native plants on plant species richness in invaded communities (Vilà et al. 2015). We compared the number of native plant species in plots dominated by a single non-native plant species with paired uninvaded control plots to assess whether the magnitude of impact was dependent on some of the major characteristics of the non-native species and/or the broad characteristics of the invaded site. As the data set accounted for 81 different species from 31 families, we also considered the influence of shared evolutionary history among species. Specifically, we used six categorical variables and the phylogeny of the non-native species as predictor variables. Three of these variables were non-native species descriptors: life form (tree, shrub, perennial forb, annual forb, perennial grass, and annual grass); presence of either clonality or vegetative reproduction (yes or no); and ability to fix nitrogen (yes or no). The three other variables were related to the type of invaded ecosystem (forest, shrubland, grassland, old field, ruderal, desert, riparian, coastal, wetland); biogeographic region (temperate, Mediterranean, tropical, sub-tropical, arid, and semiarid); and insularity (whether the study was conducted on an island or not).

On average we found that non-native plants reduced plant species richness by 20.5%. Of the six categorical variables assessed, clonal growth and N-fixing were the only ones influencing the magnitude of the impact. Clonal plants or plants with vegetative reproduction reduced species richness more than non-clonal plants (Box 2.1 Fig. 1).

2.2 Correlates/Traits Associated with Impact · 31

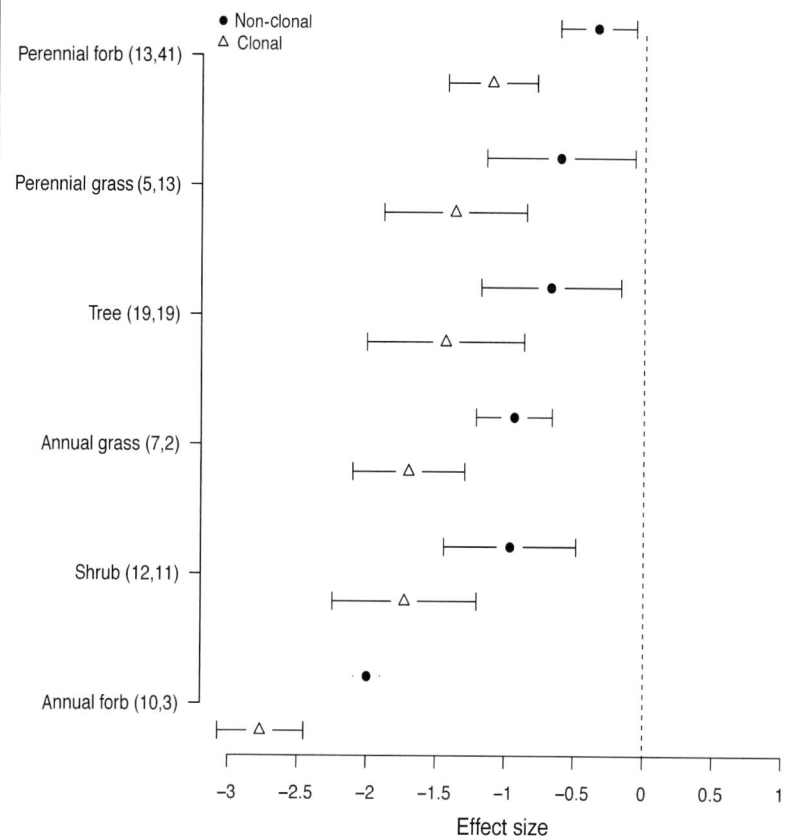

Figure 1. The impact of non-native plant species on native plant species richness. Effect size (±1.96 SE, i.e. 95% confidence interval) is computed as the log-ratio of the number of species in the invaded plot over the control plot. An effect size is significantly different from zero when its 95% confidence interval does not bracket zero. A negative effect size indicates a decrease in plant species richness. Sample sizes for non-clonal and clonal species are indicated respectively in parentheses. Reproduced from Vilà et al. (2015) with permission.

In plant invasion biology there has been a lot of emphasis on the impact of N-fixing species on N-cycling (Castro-Díez et al. 2014). However, contrary to the general wisdom, N-fixing species reduced plant species richness less than non-N-fixing species. In fact, there have been many cases of N-fixing non-native species (e.g. *Acacia* spp.) not reducing local species richness in all study sites.

The most striking result was the presence of a phylogenetic signal on the magnitude of impact. Closely related species tend to have

> impacts of comparable directionality and strength. The cause of this signal is probably due to closely related species sharing traits that might increase competitive ability. Although our study did not precisely identify these traits, our results support the use of information from closely related species to infer potential impacts of an unknown invader in risk assessments.

2.3 Understanding Mechanisms of Invasion and Impact

While such exercises in determining correlates and traits of invasiveness and impact have substantial conceptual interest, and provide a link between ecological and evolutionary theory and invasion science, it is important to keep in mind that the predictive power is still limited. As such, the broad generalisations that have emerged to date have interest for research, but the practical consequence is that either a substantial invasion risk needs to be allowed or unnecessarily restrictive regulations are enacted. If, however, we could understand the mechanisms of invasion and impact, then it might be possible to understand, predict, and effectively manage the risks.

There are several performance-related traits that are directly associated with invasiveness (e.g. physiology, leaf-area allocation, shoot allocation, growth rate, size and fitness (van Kleunen, Weber, & Fischer 2010)), but trait data are not always readily available to risk assessors. However, in some cases a single mechanism can be identified that acts to prevent an invasion (e.g. the presence of herbivores or the lack of a sustaining resource). Once the limitation is lifted, the invasion occurs. For example, certain leguminous species failed to set seed before the introduction of buzz-pollinators to New Zealand (Richardson et al. 2000a). The dynamics of a number of plant invasions involve a significant **lag phase** between introduction and invasion, sometimes extending to many decades (Box 2.2; Kowarik 1995; Crooks 2005; Aikio, Duncan, & Hulme 2010b). Any delay presents an opportunity for very effective intervention at a stage when a species is highly restricted in space, but equally there is a risk that resources might be expended to manage a plant that might otherwise never become a significant problem (e.g. the plant population is not in a lag phase, it simply will never expand widely). How to prioritise these species is essentially a risk-management issue (Chapter 4).

Box 2.2 *Lag Phases: Theory, Data, and Practical Implications (Petr Pyšek)*

Why Do Lag Phases Occur?

Lag times in invasion biology typically refer to either the delayed onset or relatively slow rate of an invasion (Crooks 2011). Many studies have found a period of slow initial population increase, followed by a phase of rapid range expansion (exponential phase), and a third phase of little or no area expansion (see Pyšek & Hulme 2005 for a review). Lag phases of 40–200 years have been documented for herbaceous species (Pyšek & Prach 1993; Crooks 2011) and woody plants (Kowarik 1995). Factors that determine the length of the lag phase can be divided into three groups (Pyšek & Hulme 2005):

(1) The genotypic hypothesis suggests that a lag phase reflects the time required for the development of genotypes that have an ability to spread rapidly (Hobbs & Humphries 1995), and predicts that the length of the lag phase is proportional to generation time, and that long-distance dispersal is an intrinsic attribute of the species concerned. While there are data supporting the first prediction (Kowarik 1995), long-distance dispersal events appear to be largely driven by extrinsic factors (Higgins, Nathan, & Cain 2003). Similarly, the introduction of a new genotype might stimulate an invasion.

(2) The demographic hypothesis advocates that soon after introduction, alien populations expand slowly at their margins by short-distance dispersal and spread is limited by the local availability of suitable habitat (Cousens & Mortimer 1995). The rapid spread associated with exponential increase is initiated by long-distance dispersal that establishes new satellite populations in suitable habitats. Linked to the species demography is a possible effect of life-history on the duration of the lag phase (Pyšek & Prach 1993). These authors reconstructed over 200 years of invasion history of only four plant species, but the pattern was quite striking: clonal perennials not reproducing by seed in the invaded region (knotweeds, *Fallopia*) had to establish in more sites before the start of the exponential phase of invasion than prolific seed producers (*Impatiens glandulifera*, *Heracleum mantegazzianum*) that entered the exponential phase of spread after having established in only a few localities. Moreover, the spread of the latter two during the exponential phase was about twice as fast as that of knotweeds.

(3) The extrinsic hypothesis proposes that exponential population increase occurs only after some extrinsic conditions have improved (Sakai et al. 2001), such as changes in soil disturbance, nutrient enrichment, climatic changes, arrival of dispersal vectors or pollinators, and intra-specific interactions. For example, anthropogenic disturbance, partly related to WWII, promoted the spread of Oxford ragwort (*Senecio squalidus*) in Europe, a weed that escaped from a botanic garden as early as the 1700s (Crooks 2011).

Pyšek and Hulme (2005) point out that although all three hypotheses are plausible, and not mutually exclusive, the lag phase might be no more than an artefact, because it is notoriously difficult to distinguish statistically between a single exponential phase of increase and one that has both a lag and exponential phase (Cousens & Mortimer 1995). Also, the lag phase depends on the scale of observation; invasions are discontinuous in time and space and comprise both local population expansion as well as new introductions, hence what appears to be a lag might actually conform to a constant exponential expansion rate when viewed from a coarser spatial scale (Pyšek & Hulme 2005).

Evidence for the Existence of Lag Phases

In general, most data refer to case studies of individual species, e.g. *Spartina anglica* in the UK (lag phase of 40 years), *Schinus terebinthifolius* in Florida (100 years), or *Spartina alterniflora* in Washington (100 years) (Crooks 2011: table 1). Attempts to identify general patterns that are valid across alien floras are much scarcer. In a classical study that was the first to estimate the duration of lag phases for a number of species, Kowarik (1995) gathered data for 184 woody plants cultivated in Brandenburg, Germany, and found that on average the lag phase lasted 147 years (Box 2.2 Fig. 1) and that the most successful invaders are not necessarily those that invade first.

Williamson et al. (2005) analysed the character of curves describing increases in the number of localities over time in the Czech Republic, central Europe. About one-third of the 63 species analysed showed lag phases, and many species exhibited deceleration at the later phase of invasion. The area occupied by a species at the regional scale of the country doubled in about ten years, and the invading species spread 2 km per year. Those species that show a lag started to spread about 25 years earlier than those that do not, and newer introductions spread

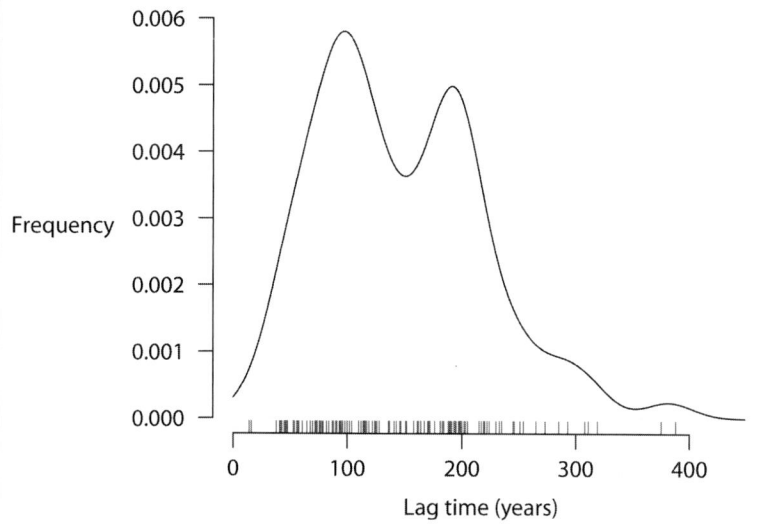

Figure 1. The distribution of invasion lag times for woody species introduced to Brandenburg, Germany. The data shown are for 163 woody species where the lag time is the number of years between the first record in cultivation and the date at which the species was recorded in the wild (i.e. had escaped cultivation) in Brandenburg (in some cases more specifically Berlin). The mode lag time for shrubs (first peak) is about 100 years less than for trees (second peak). The graph was drawn using the *density* function in R, with a bandwidth set to 20 years. Data from Kowarik (1995).

faster because landscapes have changed and become more suitable for fast spread of alien plants.

What Do Lag Phases Mean for Management?

One important fact associated with the existence of lag phases is that the number of invading species is bound to increase, even if further introductions were to cease (Kowarik 1995); the species are there waiting for their lag times to be over. Yet, the existence of lag phases represents an opportunity for early warning and rapid response because species that start to spread exponentially immediately after their arrival to a new region are much more difficult to eradicate or contain than those that start with fewer populations and that 'take their time'. Recognition of lags emphasises that decades or even centuries might pass before species undergo ecological explosions; many species did nothing for decades or even centuries, only then to become problems. Lags imply that in many instances there might be relatively long windows of opportunity when invasive populations are small and confined (Crooks 2011).

In the above-mentioned study of Williamson et al. (2005), the majority of species showed no lag phase. Besides issues that make it difficult to detect the lag phase statistically (Cousens & Mortimer 1995), it seems likely that in many contemporary invasions there are indeed no lag phases. One reason is that as the number of introduced species has been rapidly accumulating (Lambdon et al. 2008; Hulme et al. 2009), increasingly more species are being introduced from areas in which they are not native, often in close proximity to the region of study, meaning that they are coming already well-adapted to local environment and biotic interaction – their lag phase occurred long ago, upon the first introduction. While the region of central Europe was the first site of introduction for some early newcomers in the eighteenth and nineteenth centuries, later arrivals (such as those after WWII) were often species that might have gone through their lag phases in other parts of Europe. Ability to detect a lag phase therefore depends on the period of invasion (increasingly less likely to play a role in modern mainland landscapes) and on the scale (more likely to be observed at large scales, such as whole continents). The question thus arises, are we going to observe, at least in heavily invaded continents such as Europe, 'classical' lag phases less often than in the early days of invasions?

A related line of research takes the position that all species could become invasive somewhere; if there are no invasions, then potentially limiting factors should be tested in turn to see which factor explains the lack of spread. For example, Bufford and Daehler (2014) looked at horticultural plants that have been present in Hawai'i for a significant period and planted in reasonable numbers. As such, introduction dynamics could be ruled out as the cause of a lack of naturalisation. By performing a set of carefully designed tests, they determined which reproductive factors were limiting naturalisation, e.g. *Hemigraphis alternata* has low pollen viability, while seed set in *Allamanda cathartica* appears to be limited by a lack of compatible mates. From this it is possible to draw conclusions for risk assessment, e.g. the introduction of new types of *Allamanda cathartica* might provide compatible mates and so result in the production of viable seed, potentially leading to naturalisation. Autecological studies of this type, while sometimes time-consuming and costly to conduct, can provide a vital contribution to post-border weed risk assessments, where the aim is to understand the key drivers that determine current patterns of abundance and extent, and under which conditions there is likely to be a significant change in the future. By a process of elimination and induction, the various factors (e.g. propagule pressure, residence time, reproductive failures, competition, and disturbance) can be assessed and hypotheses as to the driving mechanisms refined.

It is also important to note that while the outcome of plant introductions is often considered in binary terms (invasive or non-invasive), most species will naturalise at some sites but not at others, and the likelihood and consequences of introductions are influenced by the introduction dynamics. It is therefore important to understand the context-specific mechanisms underlying naturalisation and invasion (Moodley *et al.* 2014). For example, *Banksia ericifolia* was, based on biological traits, predicted to be the *Banksia* species most likely to invade the fynbos of South Africa (Richardson, Cowling, & Le Maitre 1990). Despite decades of cultivation for flower production, the first record of an invasive population was made only in 2008. This appears to be because most populations were cultivated and so protected from fire (as a largely serotinous species, *B. ericifolia* releases most of its seed only after fire). When plants were subject to the natural fire regime in the area (due to a change in land usage), large numbers of seeds were released, and plants quickly established, resulting in an invasion. Therefore the initial trait-based analysis was correct in predicting that the species had invasive potential. By determining the necessary conditions (fire), fairly simple rules could be developed to limit invasion risk (i.e. don't let plants burn) (Geerts *et al.* 2013). Invasion science will continue to benefit from a species-by-species, population-by-population approach. This might appear daunting, but equally should mean that the risks involved can be managed appropriately.

2.4 Risk Assessment

Although the current state of understanding is far from comprehensive, considerations of correlates, traits, and mechanisms of invasiveness and impact somehow need to be translated into policy and action if they are to be of value. This is conducted in a formal process of risk analysis. A risk analysis is composed of a risk identification, risk assessment, risk management and risk communication. Risk assessment addresses the likelihood of a damaging event occurring (expressed as a probability) and the consequences if the event occurs (expressed in terms of magnitude, with negative consequences generally known as hazards). In the context of invasive plants, 'likelihood' refers to the probability that an introduced plant will establish and spread, while 'hazards' encompass all of the negative impacts that could result as a consequence of its spread (Daehler & Virtue 2010). Risk management is used to identify and evaluate potential options that can be applied to address the risk, while risk communication focuses on how to present the findings and recommendations to stakeholders. We discuss risk management and risk communication in later chapters (in particular Chapters 4 and 8 respectively); here, we focus on risk assessments.

The basic elements of all invasive plant risk assessments were outlined in a scheme developed by Leung *et al.* (2012), named 'TEASI'. The TEASI scheme (standing for *t*ransport, *e*stablishment, *a*bundance, *s*pread, and *i*mpact) is abstract in nature, but provides the framework for developing risk assessments that will map onto the set of real processes underlying invasions (Leung *et al.* 2013). However, to date, few schemes consider all five components (transport, establishment, abundance, spread, and impact); they usually focus on a species perspective, are qualitative, and are specific to a particular goal (Kumschick & Richardson 2013).

One of the most widely adopted pre-border risk assessment systems was implemented in Australia in 1997 (Pheloung 2001; Weber *et al.* 2009; Gordon *et al.* 2010). This semi-quantitative scoring method (often termed the Australian Weed Risk Assessment (AWRA) system) is usually used to arrive at a binary decision either to allow entry or prohibit importation into a country, although a third option (requires further evaluation) is available. The AWRA system has been adapted and utilised elsewhere (Pheloung 2001; Gordon *et al.* 2008; Nishida *et al.* 2009), sometimes with the addition of a secondary screening tool to reduce the number of cases requiring further evaluation (Daehler *et al.* 2004). The AWRA system, and its modifications, have proven sufficiently sensitive (i.e. accurately identifying potentially harmful plants) and specific (not designating benign plants as potentially harmful) to be useful as a screening tool (Gordon *et al.* 2008).

Risk assessment procedures have also been developed which focus on specific environments, for example aquatic plant invasions (Champion *et al.* 2010). While these provide useful standard ways of summarising information on species, they still have some fundamental problems – for example, they do not explicitly link to mathematical probabilities. In general, although weed risk assessment is no longer particularly young as a discipline, a number of important matters remain to be addressed properly. Hulme (2012) discusses a few: the development of objective measures for hazards; difficulties associated with predicting plant behaviour in complex systems; quantification of uncertainty and variability; and the often-unappreciated cognitive biases of experts.

Once a plant has breached a border and has been detected in the environment, it becomes a possible candidate for post-border weed risk assessment. As before, the aim is to predict the behaviour of species in the new environment, but in the context of the current known distribution and likely modes of dispersal from these sites. While in many cases there is no difference between pre- and post-border risk assessments, post-border risk assessments should employ post-introduction models that rely heavily on geographical information systems (Křivánek & Pyšek 2006). The primary objective in this context is to identify when risk management options need to be weighed up, i.e. to determine if the species is an appropriate target for eradication or containment. The process of risk management will be addressed further in Chapter 4, but, as a basic guide, risk analysis should be conducted for any plant incursion that shows signs of spread, or if species are known to cause problems elsewhere in the world, and only if resources permit should other naturalised species be considered. Any trait-based assessments of impact, as opposed to assessments based upon extrapolation from behaviour elsewhere, will be associated with a high degree of uncertainty, because an attempt is being made to predict a phenomenon that has a very low base rate (i.e. a small proportion of the total number of introductions will go on to become problem species).

At the scale that concerns land managers, impact is a function of both the *per capita* effects that arise from particular plant traits and the cumulative abundance of a species relative to others with which it co-occurs (Drenovsky *et al*. 2012). Clearly, the types and magnitudes of an introduced plant's potential impact are of most concern. Owing to the limited availability of resources, only the plants considered to have the highest potential impacts are likely to be targeted for management. Future impact will be a function of an introduced plant's potential distribution (taking into account the spatial distribution of suitable habitat) and its aggregated impacts across this distribution. From a risk-assessment perspective, not only is the native range important, but it is also important to assess how rapidly a species might spread (Lindgren 2012). In the following sections we discuss efforts made to predict potential distributions and to estimate dispersal. These are often implemented in terms of mapping risk (Chapter 3).

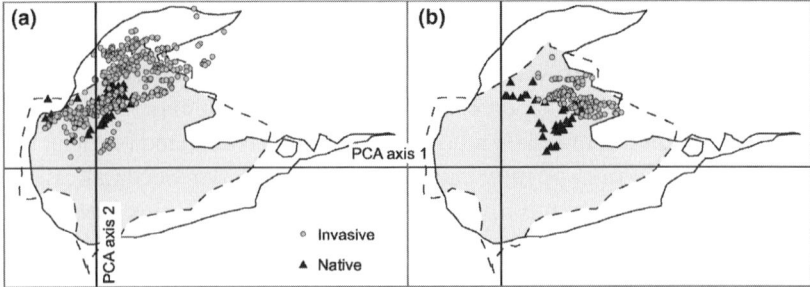

Figure 2.4. Principal component analysis (PCA) of the climate space occupied by (a) boneseed (*Chrysanthemoides monilifera* ssp. *monilifera*) and (b) bitou bush (*C. monilifera* ssp. *rotundata*) in their southern African and alien (Australia, New Zealand) ranges on the basis of six climate variables. The lines show the alpha hulls of the climate space available in Africa (outlined with dashed line) and Australia and New Zealand (solid line). Climate space common to both the native and alien ranges is coloured grey. In its alien range, boneseed occupies a broader climatic niche than it does in its native range, having moved into climate space that is unavailable in southern Africa. On the other hand, alien bitou bush populations have shifted into climate space that is not novel in the sense that it occurs in southern Africa but apparently is not occupied there. Reproduced from Beaumont *et al.* (2014) with permission.

2.4.1 Predicting Potential Distribution

An invasive plant's potential distribution corresponds to the total area that would be at risk if it were to be introduced to all locations. The simplest approach to determining potential distribution is to match the known distribution of the invasive plant with similar climatic zones (using, for example, maps of Köppen's (1936) climate classification available at www.climond.org). A slightly more sophisticated method is to use a correlative model to project the plant's native distribution onto the introduced range. Such models generally provide a useful approximation of the potential distribution of the plant, which can be improved by including observations related to the plant's distribution beyond its native range (Broennimann & Guisan 2008; Jiménez-Valverde *et al.* 2011). One complication is that climate change is likely to cause significant modifications of the distributional ranges of many species, and so future predictions need to be treated with caution. Moreover, when species are introduced beyond their native ranges they might encounter novel climates (Fig. 2.4) and might either demonstrate pre-adaptation to these or be able to expand their climatic tolerances when freed from their natural enemies.

In developing spatial predictive models there are multiple steps in the process where risk analysts and practitioners are required to make informed decisions (Box 2.3). These include selecting the most appropriate model, interpolating the results, evaluating data quality, evaluating model performance, validating the model, testing the model, and selecting the most important constraining variables. Models only perform as well as the selected predictive variables. Variables that generally determine potential distributions are abiotic conditions in the local environment (e.g. climate), biotic interactions with other species (e.g. competition and predation), and the ability of a species to disperse (Hirzel & Lay 2008). The potential distribution of an invasive plant is then the novel area where abiotic and biotic conditions are favourable and where it is able to disperse or move into.

Box 2.3 *Species Distribution Models (Jane Elith)*

Models for predicting the distribution of invasive species are used in two main situations. One is when maps of potential distribution are required, often before a species has entered a region. The other, not addressed further here, is for understanding the current distribution of species as they spread in the invaded range (Václavík & Meentemeyer 2009). The basic building block for many models used for both purposes are correlative species distribution models (Elith (in press); also known as ecological niche models) that relate records of the species occurrence or abundance to a suite of predictor variables. Modelling methods vary depending on the type of species data available (presence-only, presence–absence, or abundance), and include envelope-style methods, distance-based methods, and statistical and machine-learning methods, including various regression models.

Accurate predictions of potential distributions require knowledge of the fundamental niche of the species (i.e. that set of resources and conditions that allow the species to exist indefinitely (Colwell & Rangel 2009)), but this is almost always unknown and in most cases difficult to estimate. Instead, species distribution models (SDMs) are fitted to records of species occurrence in native ranges and perhaps from invaded ranges where the species has persisted (Jiménez-Valverde et al. 2011). The main problem is that environments in those regions might not represent the full suite of conditions in which the species can survive, so any model built on these records might reflect relationships in the native range but will not include the species' response to environments outside those of the known occupied regions. Biotic interactions and dispersal limitations have also impacted observations from the realised niches, and in most cases data are not available to allow these impacts to be quantified. Therefore impacts of these processes on the known records will be implicitly included in the fitted SDM, and will be assumed comparable in new regions.

Keeping these limitations in mind, good practice in using SDMs to model potential distributions include the following: (1) Gathering as much information on the species occurrence as possible. The best-quality data are records of abundance or presence–absence, since these produce absolute predictions of abundance or probability of presence. However, presence-only records (records of presence without accompanying information on sites surveyed but found unoccupied) are most common but more challenging to use because of difficulties in dealing with bias, and with the relative scaling of the output (predictions are not absolute probabilities of occurrence (Guillera-Arroita et al. 2015)) (2) Selecting those candidate predictor variables most likely to influence the distribution of the species, including representations of climate, substrate, and topography, depending on the scale at which models and predictions are required (global vs regional vs local). (3) Understanding the modelling method, and selecting settings relevant to the task. For instance, some models by default allow very complex fitted models, but simpler models might be more relevant for giving predictions for new regions (Merow et al. 2014). Many methods model presence-only data by comparing them with background (or pseudo-absence) data, and the number of records and geographic extent of the sample can strongly impact the result (Renner et al. 2015). (4) Quantifying which areas in the range of interest represent novel climate space compared with the native range (Elith, Kearney, & Phillips 2010; Zurell, Elith, & Schroeder 2012; Mesgaran, Cousens, & Webber 2014). In these the model needs to extrapolate beyond the training data, and might be unreliable (see Elith (in press) for further details). Beyond correlative models, mechanistic (biophysical) models that capture those processes contributing to survival and fecundity can also be used to predict distributions and in many ways are more relevant to invasive species because they directly attempt to model the fundamental niche (Kearney & Porter 2009; Morin & Thuiller 2009). However, they are difficult to parameterise so applications are restricted.

How might modelling approaches change in the future? Given the ongoing effort in digitising species records, making them available online, and setting up systems for ongoing refinement and correction of the records, it is likely that gradually more complete records will become available. These will be most valuable if effort is directed to countries so far poorly represented (Amano & Sutherland 2013) and to making available both presence-only and presence–absence data. Modelling methods are emerging that combine data types (Dorazio 2014; Fithian *et al.* 2015), and these provide a powerful means for dealing with biases common to presence-only data. Similarly, there is global interest in developing predictors relevant to biodiversity, leading to increased availability of global coverage for ecologically relevant variables, including climate extremes.

It is likely that the limitations of using correlative methods for predicting potential distributions will be increasingly clear and well defined. While SDMs might be the only practical tool in many situations, predicted distributions will always be to some extent uncertain despite improved data because they rely on records from the realised (rather than the fundamental) niche, and because the true causal variables that would predict distributions are unknown and probably unavailable. A better appreciation of these issues should lead to further integration of insights from mechanistic models – for instance, by generating predictor variables from mechanistic models (e.g. Kearney, Isaac, & Porter 2014) that are physiologically relevant to the species, or by specifying limits to distributions based on physiological knowledge. Some facets of this approach are already available in some methods (e.g. CLIMEX (Sutherst & Maywald 1985)), but it is likely that other approaches will emerge. Open-source or free versions of code are especially valuable. Models that can be controlled as to how they extrapolate would be useful, as would more methods for identifying novel climates. Regardless of such advances, it is certain that predictions will still be error-prone, and development of methods for dealing practically with uncertainty would be valuable. The impact of uncertainty varies with the application. In some cases, users can tolerate errors in predictions on a species-by-species basis, as long as there is no consistent bias across all species. This is true, for instance, where predictions are used for allocating costs of surveillance and containment to states within a country. For example, one might want to clearly delineate places where the species will not survive. Advances in dealing with uncertainty will allow considered decisions about priorities.

Species distribution models are now commonly used for predicting the potential range of invasive plants and should be considered as an important component in the biosecurity toolbox (Lindgren 2012). Predicting the potential distribution of a pest represents a shift in the biosecurity paradigm from a reactive to proactive approach. Spatial and temporal characterisations of risk or models of potential distribution are required in order to develop strategies to respond to an invasive plant (Venette et al. 2010). Models of potential distribution are very useful in preparing response strategies for invasive plants, as limited resources can then be prioritised (Waage & Mumford 2007). Spatial predictive models can be used by decision makers to inform regulations as well as to identify high-risk areas for surveillance, leading to efficient allocation of survey resources. Predictive spatial modelling is not without its challenges, however. These include data quality (i.e. non-validated point occurrence data, inaccurate and biased data), accounting for biotic interactions, spatial autocorrelation, estimating error (omission and commission), and addressing dispersal events.

Where available, information on failed invasions can be a useful input for predicting potential distributions of invasive species within an area of interest. If data on failed naturalisations/invasions exist, and failures can be attributed to deterministic causes, it will be possible to subtract unsuitable regions from the potential area and obtain a more accurate prediction (Zenni & Nuñez 2013). For example, data on the success or failure of forestry trials provide a good test of, as well as corroborative evidence for, species distribution models (Motloung et al. 2014).

Depending upon how species distribution models operate, reference locations that are outside the 'normal' distribution of the plant (i.e. outliers) might have a major influence on model projections. Some of these arise from errors in herbarium data, and there are some relatively straightforward techniques to identify and resolve them (e.g. latitudes and longitudes might be transposed (Robertson et al. 2016), but if resources permit, using ground-truthed data would be preferable. In other cases it is important to understand the species' life-history and ecology, and in particular its habitat preferences in its native and introduced range. For example, if species are dependent on micro-climatic conditions (e.g. riparian corridors and swamps), species distribution models based on broad-scale climatic conditions can be wildly inaccurate. Similarly, invasive plants can grow outside their modelled range in irrigated farming systems. The effect of irrigation can be modelled separately using global data sets that capture the extent of irrigation practices (Siebert et al. 2005). For aquatic plants, rainfall is largely irrelevant, and temperature-based estimates should be supplemented with geographic data on the occurrence of water bodies to obtain a better estimate of the area at risk of invasion, again with the proviso that broad-scale air temperatures might not accurately reflect the temperature experienced on or in a water body. Further refinement based on hydrological properties is possible, with key factors including flow rates and salinity, water depth and clarity (Champion & Clayton 2001).

Soil preferences can be used in conjunction with climate-based modelling to refine potential distributions of plant species, but caution should again be exercised. Soil types comprise many different attributes (e.g. texture, pH, nutrients) and it might be unclear which attributes are range-limiting. Furthermore, because characterisation of soils is complex and expensive, available soil mapping for many regions of the world is nowhere near as comprehensive as is mapping for climates. In general, the resolution of the cells used in the predictive analysis need to relate to the spatial distribution of the underlying processes. But even if data are available, they might be exorbitantly expensive.

Invasive plants are likely to invade and cause impact in certain land uses and ecosystems. When maps of these land uses are available these can be overlaid with the climatic and soil tolerances to further refine potential distribution. Within the land uses there might be capacity for further refinement by selecting particular vegetation types at risk (e.g. specific crops or vegetation communities). This concept is developed in Section 3.2.

2.4.2 Predicting Dispersal

How quickly an introduced plant spreads will determine the rate at which the damage it causes will accumulate, and as such it is important to predict dispersal. For example, given equal hazards posed by two different introduced plants, the species with the higher rate of spread will be more important to target because the immediate benefits associated with successful intervention are greater. This point was expressed in economic terms by Harris and Timmins (2009, p.6): 'There is less benefit to be had from early detection or control of species that spread slowly, because any increase in control costs over time is offset by the change in the value of money over time (discount rate).'

An invasive plant's rate of spread is determined by its population growth and its dispersal characteristics, but recent work (Bullock, Pywell, & Coulson-Phillips 2008; Coutts *et al.* 2011) has supported the notion that spread is driven primarily by dispersal (particularly human-aided),

Table 2.2. *Invasive plants will vary in their potential for effective long-distance dispersal (i.e. dispersal leading to plant establishment). Human-mediated dispersal can particularly enhance invasiveness, increasing the likelihood of rapid spread. The probability of effective long-distance dispersal is also affected by the propagule pressure and by where plants occur in the landscape relative to potential dispersal vectors.*

Vector	Potential for effective long-distance dispersal	Key dependencies
Wind	+	Terminal velocity
Water	++	Fruit type; degree of buoyancy
Wild animal	++	External/internal carriage, with or without flight
Human-mediated	+++	Domestic animals: External/internal carriage; transport distance
	+++	Machinery/product: degree and pattern of contamination; transport distance

with the demographic features associated with population growth being of secondary importance. With regard to spread modelling, Rejmánek et al. (2005, p. 128) concluded 'that the type of dispersal can have a qualitative impact upon the spread predictions, whereas the population growth terms have only a quantitative influence'. Dispersal ability depends on the number of dispersal modes for a plant, the frequency of a mode's occurrence (which might vary for different plants and land uses) and the dispersal distances achieved via different **vectors** (Table 2.2).

The spread rate is related to propagule pressure. Generally, most propagules come to rest relatively close to the parent plant as a result of short-distance dispersal events; very few are deposited remotely through long-distance dispersal events. Some dispersal mechanisms (e.g. dispersal by vertebrates, water, or wind) tend to generate considerably more long-distance dispersal events than others (e.g. dispersal by ants, fruit 'explosion', rain splash or gravity). However, substantial dispersal distances are often achieved only through co-option of human activities (e.g. seed contaminants can be transported globally, and the horticultural sale of invasives in new countries lead to widespread dispersal). Human-mediated dispersal can cause large numbers of propagules to be frequently moved large distances to areas suitable for establishment, hence increasing spread rates.

Plants are commonly spread by more than one dispersal vector, the combination of vectors being termed a 'vector suite' by Panetta and Cacho (2012). If a number of dispersal vectors are involved, the vector that contributes most to long-distance dispersal will most influence the rate of spread. In some cases the vectors involved for a newly detected invasive plant might be known (or at least suspected on the basis of field observation). However, Thomson et al. (2010) have argued that the dispersal capabilities of most plant species are unknown. These authors used linear mixed models with basic life-history and ecological traits to predict dispersal mechanisms for species for which such information is limited. They created sets of models for six dispersal categories (unassisted, wind, water, ant, vertebrate ingestion, and vertebrate attachment dispersal mechanisms). The main intrinsic predictor variables for dispersal mechanisms were growth form (tree, shrub, herb, or climber) and seed mass (categorised as $<0.1, 0.1–100,$ and >100 mg), with vegetation type being a significant extrinsic predictor variable. Their models predicted dispersal mechanisms for Australian and Californian species equally well (~70% correct in both cases), but didn't perform as well for Swiss species (~50% correct). Subsequent work showed that both mean and maximum seed dispersal distances were more strongly correlated with plant height than with seed mass (Thomson et al. 2011).

The creation of new populations is a risky business (from a plant's perspective), and unless relatively large numbers of propagules are involved, the probability of establishment and persistence might be very low (Minton & Mack 2010). For this reason, rates of spread are likely to be highest where multiple propagules are deposited simultaneously and/or repeatedly. This is often likely to occur where humans and their agents are involved. Dispersal by humans might be either intentional (as where potentially invasive species are grown as crops and pasture or used for landscaping and gardening – providing propagule sources for new populations) or accidental (unintentional). In the latter case, a large number of mechanisms might be involved, including dispersal via humans directly (e.g. species whose propagules can attach to clothing and footwear) or indirectly via attachment to vehicles (e.g. cars, trucks, earth-moving and farm machinery and boats). Other types of human-mediated accidental dispersal occur via contamination of commodities (e.g. ornamental plants, hay, grains, soil, gravel, mulch or turf, or by-products like waste from stockfeed manufacturers and textile mills).

There is relatively limited information available on the distances achieved through different modes of dispersal (Vittoz & Engler 2007;

Thomson *et al.* 2010). We know of no examples where the proportions of propagules dispersed via different vectors have been quantified for a plant species dispersed by multiple vectors. Both of these factors are major contributors to the uncertainty associated with analyses of invasiveness. Where more than one mode is involved, it is extremely difficult (if not impossible) to quantify the relative importance of each (Panetta 2012), but for the purpose of risk management it is best to focus on the mode capable of facilitating the most rapid spread. For a discussion on the pivotal role of long-distance dispersal in plant invasions, see Trakhtenbrot *et al.* (2005).

From theoretical and observational lines of evidence it is clear that the influence of humans on dispersal (via introduction dynamics) can have a greater influence on the abundance and extent of an invasive species than any inherent biological traits (Donaldson *et al.* 2014). So it is unsurprising that, as far as quantitative estimates for potential spread rate are concerned, plant traits tend not to be strongly correlated with observed spread rates (Gravuer *et al.* 2008). Human-mediated dispersal can be responsible for the rapid spread of many species that exhibit no particular adaptations (Benvenuti 2007). For the genus *Trifolium*, Gravuer *et al.* (2008) found that different species have become widespread in New Zealand via one of two pathways: (1) economically useful species introduced early by Europeans and planted widely; and (2) species that were unintentionally introduced, with some spreading rapidly via pathways linked to human activities. Species in each of these groups shared different biological traits. For example, the intentionally introduced species were predominantly self-incompatible perennials, whereas the unintentionally introduced species were predominantly small-seeded annuals. Gravuer *et al.* (2008) argued that these two pathways favoured species with very different trait suites, and helped to explain why few biological traits were associated with relative rate of spread in this genus, suggesting that biological traits might play a larger role in influencing rate of spread in natural habitats.

Quantitative estimates of potential spread rate for an invasive plant can also be based on data gathered elsewhere in the plant's invasive range. There are obvious limitations on the inferences that can be drawn from such extrapolation, but some information is generally better than none. Here it will be critical to anchor observations in both time and space, i.e. to be able both to determine the ages of newly detected populations and to link them to their sources, often difficult undertakings (Panetta 2012).

A range of sophisticated models has been developed to mimic and/or investigate the spread of plant species, both in simulated and real landscapes (Bullock, Pywell, & Coulson-Phillips 2008; Fox *et al.* 2009; Smolik *et al.* 2010). Depending upon their complexity, these models require varying amounts of species-specific information. In the absence of first-hand observations on spread, any of a number of simple models might be used to generate spread scenarios, again depending upon the information available. A dispersal model combining a landscape-explicit, natural dispersal kernel (based upon most likely vector or vectors) with routes of human-mediated transport (where relevant) could be used to model spread (Wilson *et al.* 2014). However, if decisions are made at a stage where early intervention is feasible, there will likely be insufficient information to parameterise a population growth rate model. In this case it might be best to develop a few growth rate scenarios and either base estimates on the scenario considered most likely or take a highly precautionary approach and assume high rates of population growth. Again, this should be undertaken only for the species that are considered to pose the greatest threats, and much more work is needed to test the models in terms of their contribution to on-ground management.

Recently the European Union project PRATIQUE (Enhancements of Pest Risk Analysis Techniques) developed and tested quantitative models for estimating and mapping the spread of pests which are largely lacking in current pest risk analysis schemes (Kehlenbeck *et al.* 2012). The general spread module consisted of four models, each providing different perspectives on spread based on logistic growth, radial range expansion, population growth and dispersal kernels. Accordingly, each model has different data requirements, with the choice of model being determined largely by data availability. Outputs of all the models are a series of maps illustrating the spread of a pest through time. The spread module is linked to the European Plant Protection Organizations' Decision-support scheme for pest risk analysis.

2.4.3 Predicting Impact

For investment in expensive mitigation efforts to be justified, there should be a reasonable and quantified expectation that the invasive plant will cause substantial negative impacts. A number of authors have suggested that given a suitable climate in the area at risk, a history of weediness is the most reliable predictor of whether a plant will establish, spread, and cause damage (Panetta 1993; Lockwood *et al.* 2001; Hulme 2012).

And in practice, 'weed elsewhere' is the most important plant attribute taken into consideration when deciding whether to manage a particular species through a coordinated programme. The vast majority, if not all, of the documented cases of attempted eradication or containment involve plants that have been targeted because they are known to have caused significant impact elsewhere. This is not surprising, since those responsible for assessing and managing risks associated with invasive species need to allocate scarce resources between current and potential invasions 'through a process of balancing disparate risks, costs, and benefits that are nonuniformly distributed under conditions of scientific uncertainty' (Andersen et al. 2004, p.789). It is not surprising that *Caulerpa taxifolia* was of immediate concern in California, given its history of impact, but pom-pom weed was relatively ignored in South Africa because it was the first invasion by the species anywhere in the world.

Viewed from this perspective, the challenge faced is twofold. We must consider how to assess the impact posed by incursions of recognised pest plants, but we need also to address the issue of species that have no previous history of invasion or impact (i.e. plants with 'non-informative priors', *sensu* Diez, Hulme, and Duncan (2012)). This is no minor issue. Williams, Nicol, and Newfield (2001) showed that of the species naturalised in New Zealand post-1940, 20% had no history of weediness elsewhere. We suspect this is an area where theory needs to be tested in practice more, and practical observations need to inform the theory. Much of the theory is prescriptive but poorly, if at all, tested.

2.5 Recommendations

When assessing the likelihood of species naturalising, invading, and causing impacts, there are some rules of thumb that can be followed, e.g. invasive plants tend to have large native ranges. But there are many important exceptions and it is important to look beyond biological species traits and consider traits of the recipient environments and the introduction dynamics. As a result, any assessment of the risks of invasiveness and impact should be treated with a large degree of caution unless the underlying biological mechanisms are understood. It is worthwhile analysing the invasion process in detail for specific groups as this can often provide valuable insights into how risks can be avoided or mitigated. Risk assessments should include predictions of potential distributions, dispersal routes, and potential impacts. Given the many pitfalls in risk assessment, experts should be consulted, either directly or indirectly through the peer-review publication process. It is vitally important that assessments are openly available so lessons can be shared.

Processes	Information requirements	Deliverables
Improve understanding of mechanisms of invasion and impact	• Lists of species at different stages of the introduction–naturalisation–invasion continuum • Evidence of invasion failures as well as successes • Experimental or field observations on potentially limiting factors	• List of traits linked to invasiveness and impact • Context-specific quantification of risk
Identify and assess risks	• Introduction pathways • Dispersal pathways within a range • Climate suitability • Information on invasiveness and impact elsewhere • Initial model for potential distribution • Qualitative estimates of potential impacts	• Prohibited lists • Watch lists (e.g. Section 6.4) • Completed risk assessments (pre-border or post-border)

3 · *Detection and Delimitation*

Key questions addressed:

- Where and along which pathways are incursions likely?
- How can the detection of incursions be improved?
- How should incursions be delimited?

A person is walking in the field and comes across a small population of a plant that she hasn't seen before. What does she do? She might be out for a walk with her dogs after a long day of meetings with clients. She stops to admire the beautiful flowers and thinks that they would look nice in her garden. She might be a field ranger late for a meeting. She quickly takes a GPS locality and makes a mental note to come back later. She might be a local botanist. She recognises it as an alien and grumbles to herself – 'Something should be done about this!' Meanwhile, the plant is busy sinking its roots that bit deeper into the ground, its flowers are withering and its seeds begin drifting on the wind. This scenario is, we suspect, common around the world. Hopefully the field ranger will come back the following day or week, or the botanist will send a report to the relevant person who will take action. But equally, people might start sharing cuttings with their friends and put a new species on the path to becoming the next big invader. Often the person forgets, doesn't have time to report the alien, or it isn't clear who should do something so nothing is done.

In Chapter 2 we discussed predicting which species are likely to be invasive and cause impact, but timely detection is often the key to an effective incursion response. Here we focus on where and how we should be looking for new incursions, procedures for improving detectability, how we can ensure that information flows from those who detect an incursion to those responsible for confirming identification and determining the appropriate response, and how to go about delimiting an incursion.

3.1 Understanding Pathways and Vectors

Introduction pathways (i.e. the routes along which species are moved) and vectors (i.e. the actual mechanism used during dispersal) have profound implications for the likelihood of invasion. At a basic level, the dispersal of species should be considered in terms of the frequency of dispersal, the number of propagules per dispersal event, and information on the sources and destinations (Wilson *et al.* 2009). As discussed in Chapter 2, such factors can have a greater influence on the abundance and extent of an invasive species than any inherent biological traits (Donaldson *et al.* 2014). By understanding the introduction dynamics it is possible to significantly improve the speed of detection, and if an incursion is detected at the time of establishment, i.e. before spread has occurred and before a large seed bank has accumulated, the probability of eradication will be high and management costs minimal.

One issue is that pathways and vectors have changed and will continue to change through time. For example, in 1686 the construction of the Lez canal from the Mediterranean Sea to Montpellier enabled wool stores to operate at Port Juvenal from 1750, with wool being received from sources around the world, washed in hot water and spread over nearby fields to dry. The wool was contaminated with new weed seeds. Over time the number of newly established species increased, until around 1860 there were almost 460 alien species present, the so-called *Flora Juvenalis*. However, with changes in wool technology, the storing and washing of wool at Port Juvenal was discontinued. By 1910, ten species remained and by 1950 only six were left (Groves 1991). This is an example of unintentional introductions. The pathway was the wool trade, with the vector being contaminated fleeces. Similar cases arose at cotton and jute mills in the UK, but the reduction in the European textile industry means that this once major pathway is now unlikely to result in any significant new introductions. By contrast, the recent interest in using known invasive plants such as *Arundo donax* and *Miscanthus* spp. as feedstock for biofuel production has created a new intentional pathway for the introduction of invasive grasses (Box 6.3).

For specific pathways the risks can be identified based on past interception records. A particularly valuable information source for this has been the records of at-border interceptions of pests. Bacon, Bacher, and Aebi (2012) demonstrated the risks associated with imported commodities based on interception and country of origin data. They devised a method to guide point of entry inspections called the TVPI or the 'Trade Volume to be Inspected Per Interception' that can be used to identify high-risk consignments. Of course, extrapolating past correlations to future trends has inherent problems (e.g. trade patterns change over time), and, as for species-based approaches, risk assessments can be improved by better understanding of the underlying mechanisms.

Some of the most interesting research in this area has been conducted on the invasion risks posed to polar regions. For example, Lee and Chown (2009) carefully cleaned luggage and cargo bound to the Antarctic, and found a substantial number and diversity of plant propagules. Plant seeds were commonly found in the socks of expeditioners and on items with Velcro (although this material also tended to prevent seeds being subsequently released). Similarly, Ware *et al.* (2012) have shown that seeds can be transferred into the high-Arctic on footwear. From this and other work (e.g. Huiskes *et al.* 2014), practical guidelines and legislation have been developed to limit such introductions (e.g. http://academic.sun.ac.za/cib/video/Aliens_cleaning_video%202010.wmv) (Box 6.1).

This is a good example of colonisation pressure (Lockwood, Cassey, & Blackburn 2009), which is defined as the number of species introduced along a particular pathway. As for propagule pressure, it can serve as a null model when comparing different vectors. However, there are a couple of important considerations. First, if a pathway exists, and propagules are transferred along it, the propagules need not be deposited in the recipient environment in a way and in a place that allows for establishment. Second, biases in which species are transported by a vector can select for or against species that will in future spread and cause impact. For example, a large percentage of invasive species come from horticultural introductions. This is partly because species are often introduced to multiple sites (Fig. 3.1), and to sites that tend to be heavily modified in a way that facilitates establishment (e.g. they are watered). However, invasive species that were introduced accidentally are more likely to end up causing large impacts, potentially because such plants need to be weedy in order to establish in the first place (Pyšek, Jarošík, & Pergl 2011). Similarly, species (especially grasses and legumes) introduced to Australia to improve agricultural and livestock production were often selected on the basis of traits that predisposed them to become invasive and weedy (Paynter *et al.* 2003; Cook & Dias 2006), and so this is a pathway that is likely to introduce **ecosystem transformers** (Driscoll *et al.* 2014).

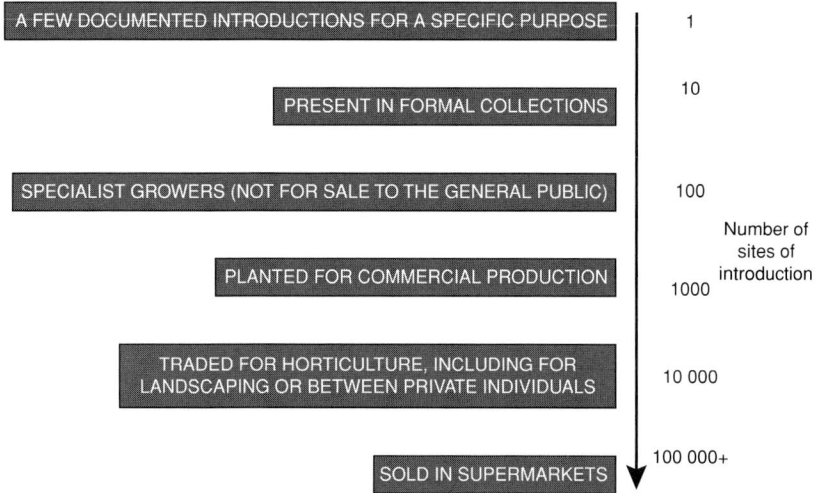

Figure 3.1. Invasive plant species differ greatly in the likely number of sites of introduction, depending on why they were introduced. This example is for deliberate introductions, but a similar trend can be determined for accidental introductions.

Historical, current, and future pathways need to be identified to assess the risks associated with introductions. This will give an indication as to which commodities need to be monitored at points of entry for unintentional (accidental) and intentional (especially deliberate illegal) introductions. However, biosecurity inspections cannot prevent all introductions in perpetuity, particularly for globally widespread invaders with a high global propagule load due to the bridgehead effect – invasive populations in one region serve as the source for colonists for other areas (Lombaert *et al.* 2010). Therefore, it is important to consider and monitor likely sites of invasion, such as along transport routes and at the final destinations of goods.

3.2 Invasion Hotspots

In addition to being introduced more often along certain pathways, invasions are also likely to occur far more often in certain parts of a country than others (i.e. invasion hotspots; for some recent analyses see del-Val *et al.* (2015) and Dodd *et al.* (2016)). As discussed previously, this is due to biases in where species are introduced (e.g. Port Juvenal in the above example), where plants spread to (both naturally and by humans), the types of species introduced, and the opportunities for invasion (e.g. at urban–wildland interfaces). For example, the relatively low incidence of invasive montane species can, in part, be explained by biases in which types of species are moved around and where species have been moved to (i.e. weeds of agriculture moved to human settlements in the lowlands), rather than any fundamental differences in species traits (Alexander *et al.* 2011).

Understanding and managing this spatial heterogeneity has been a major feature of incursion response planning. However, the introduction of plants (and their subsequent naturalisation and invasion) can be idiosyncratic. For example, results from a molecular analysis of Mexican poppy (*Argemone mexicana*) suggest a specific invasive population in a remote area of South Africa originated from a remote area of the Chihuahuan Desert. As both areas were historical centres of mining, it is possible that the introduction of Mexican poppy to South Africa was due to direct exchanges of people or equipment between the mines (K. Esler, personal communication). Similarly, one of the few populations of kudzu vine (*Pueraria montana*) in South Africa arose from rhizomes introduced in the 1920s by a farmer who had properties in both Argentina and South Africa. On visiting South America, he saw the plant being used to feed cattle, and decided to try it back home. Fortunately for South Africa, he did not share the plant with his neighbours. Kudzu vine only really started spreading with the commencement of road building in the area, remains relatively restricted, and could potentially be eradicated from southern Africa (Geerts *et al.* in review).

It would have been difficult to predict the current distribution of kudzu vine or Mexican poppy, but several approaches have been taken to attempt to derive generalisations. Analyses of the spatial patterns of invasion can be broadly grouped into three approaches: (1) analysing pathways to determine likely destinations; (2) looking for correlates of first detection records; and (3) attempting to identify areas that are likely to be susceptible to invasions.

Analysing pathways to determine likely destinations and so invasion hotspots has proven fruitful for management and research. For example, botanic gardens were a prime source of plant introductions, have a long history of being sites where new naturalisations occur, and so need to be a focus for control measures (Hulme 2011). Similarly, arboreta and trial forestry plantations often provide interesting test cases to observe the risks that particular species are likely to become invasive elsewhere. Forestry plantings have, moreover, provided useful opportunities to test ecological theory, e.g. what determines whether a species will naturalise and invade, and how quickly it is likely to do so (Daehler 2009; Dawson, Burslem, & Hulme 2009; Nuñez, Moretti, & Simberloff 2011). What such analyses have also uncovered is that current invasion patterns are due to socio-economic drivers that occurred decades or even centuries ago (Essl *et al.* 2011), in line with anecdotal evidence (e.g. of the Mexican poppy and the kudzu vine introductions to South Africa). The corollary is that future invasions will be shaped by current drivers (i.e. a lagged response).

The second approach has been to analyse first-detection localities. Herbarium or sighting records can be used to indicate 'hotspots' where naturalisations or invasions were first detected, although these data have systematic biases (Aikio, Duncan, & Hulme 2010a; Robertson, Cumming, & Erasmus 2010). Such analyses can provide broad indications of trends and where to allocate resources. For example, Huang *et al.* (2012) showed that invasive species in China were much more likely to be detected in coastal regions before they were detected in inland provinces, a pattern they argued was due to a strong correlation with trade. The more open a province was to international trade, the more introductions would occur there. Clearly, detection effort should be placed at ports of entry, but the authors also found a strong link to the funding of scientific research, suggesting a sampling bias. In general, collection data often include such biases, in part due to the spatial pattern of activities of field taxonomists.

Finally, effort can be placed in determining areas that are likely to be susceptible to invasion. In this case it is often a matter of understanding the trade-off between disturbance and invasibility – too much disturbance makes establishment difficult, but too little and there isn't a chance of establishment. This has been termed the 'weed-shaped' hole by Buckley *et al.* (2007).

By combining the likely sites of introduction, the likely dispersal pathways, and estimates of the invasibility of the landscape, spatially explicit risk maps can be provided for individual species. This can be a very powerful tool in targeting specific response efforts (Box 3.1). More broadly, such efforts can be used to determine where to put general survey effort, e.g. at urban–natural interfaces where there are water sources (Fig. 3.2). Understanding the mechanisms determining naturalisation and spread will, of course, aid the process.

(a)

SCENARIO 1

A large initial introduction associated with forestry plantings results in high propagule pressure (i.e. many seeds will be dispersing, indicated by black dots). Active management by foresters to limit spread and environmental limitations might result in few trees establishing. However, in climates that support growth, the chance of successful establishment will be increased by the proximity of the plantation to disturbed and open habitats.

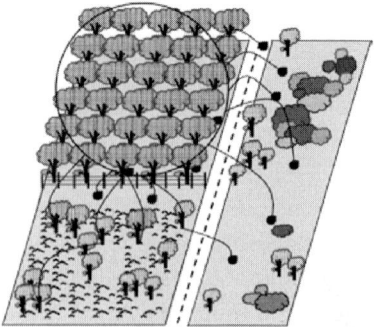

SCENARIO 2

Tree species are sometimes planted to protect infrastructure or stabilise mobile dunes. The high propagule pressure, placement within areas that are expected to support their growth, and introduction within or alongside open landscapes will promote naturalisation and spread.

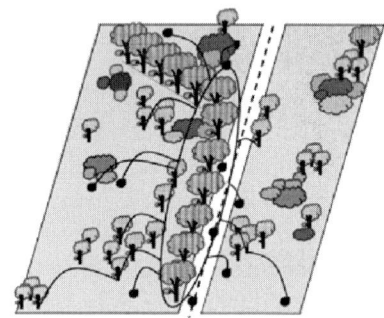

SCENARIO 3

Due to intensive cultivation, ornamental trees planted in suburban areas will be buffered from environmental fluctuations. The cultivated trees will act as long-term seed sources to the surrounding landscape. However, the combination of weak propagule pressure and limited suitable habitat in the immediate surroundings will severely limit successful recruitment. The majority of seeds will fail to establish; however, there might be highly disturbed micro-habitats, such as irrigated areas and areas of high nutrient input. Naturalised populations can form even if the broader climate is unsuitable.

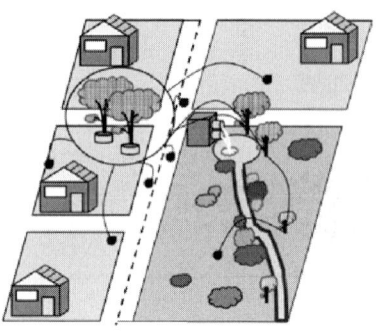

Figure 3.2. The reason for introduction can profoundly influence the opportunities available for invasion. (a) Schematic redrawn from Donaldson *et al.* (2014) with permission. (b) Scenario 1: trees introduced for forestry are often planted in large numbers in climatically suitable habitats next to disturbed and open habitats, ideal conditions for invasion. The example shown is a *Pinus contorta* invasion in Coyhaique Alto, Chile. Photo courtesy of Aníbal Pauchard. (c) Scenario 3: ornamental species are often planted in managed environments, and only establish and spread at urban–wildlife interfaces. The example shown here is *Acacia elata* in South Africa. Several large ornamental trees are visible next to the dam. Seeds have spread from these and numerous seedlings are recruiting along the dirt road and wet areas below the dam. Photo courtesy of Jason Donaldson.

3.2 Invasion Hotspots · 59

Figure 3.2. (cont.)

Box 3.1 *Risk Mapping to Underpin Post-Border Weed Management Activities (Rieks D. van Klinken & Justine V. Murray)*

Most post-border weed management activities rely heavily on predicting future impacts: what is the consequence if no actions are taken, and what actions will best minimise those impacts? There is also a strong spatial and temporal element: how are current and potential impacts distributed, and how should resources be allocated to manage these? Weed risk mapping can assist in both predicting impacts and assessing the implications of management actions.

A wide range of spatial risk modelling approaches is available. Nonetheless, their potential in guiding decision making is rarely realised, for a number of reasons. For example, with sufficient resources the scientific community could develop spatial models using available data supplemented by experimental work, but policy makers might remain disengaged. This approach can also overlook potentially important sources of knowledge. The most commonly used modelling approaches are correlative, so cannot be used to generate projections under novel environmental or management conditions. The model's spatial resolution can also be limiting, with outputs of many models such as bioclimatic ones being too coarse to be useful in guiding management programmes and policy.

One recently developed participatory modelling approach aims to overcome these issues (van Klinken, Murray, & Smith 2015). It has been developed specifically to assist decision makers in addressing a wide range of problems, such as assessing the feasibility of eradication and containment, designing large-scale management programmes, and crafting extension messages. It incorporates all available knowledge required to make spatial predictions of potential weed distribution, spread, impact and response to management. Functional understanding of the system necessary to make such predictions is captured as a hypothesis within a Bayesian network. This hypothesis is then made spatial by linking the Bayesian network with relevant environmental layers, tested and validated and altered if necessary, and then used to predict outcomes of different management interventions.

This participatory modelling approach has recently been applied to Chilean needle grass (*Nassella neesiana*). Chilean needle grass is a long-lived, unpalatable, temperate grass from South America that now forms monocultures in Australian pastures and is displacing native grasslands. One outlying population in south-eastern Queensland (eastern Australia) has been targeted for extirpation since just after its discovery in 1974. Decision makers were asking whether eradication, or even containment, was achievable, what the long-term implications of failing to respond to the invasion were, and whether long-term management options were possible.

Considerable expertise has been developed in Chilean needle grass ecology and management over the past few decades. An expert elicitation workshop was attended by Chilean needle grass researchers, land management officers and farmers to develop a model (hypothesis) for predicting Chilean needle grass habitat suitability, as well as its spread and response to management in the Queensland Murray–Darling Basin (a 700 000 km^2 region in south-eastern Queensland). The developed model identified several key drivers for establishment and persistence, including climate, land use (cultivation or pasture), ground cover, soil type, soil disturbance, and slashing (mowing) regime. Climatic factors were captured by an existing species distribution model (Bourdôt et al. 2010). The most important dispersal pathways were also captured. These were long-distance anthropogenic movement (by vehicle), short-distance movement within and between farms (by livestock and other animals and human activity), movement by flood water, and unintentional movement by slashers used to mow roadsides. Independent field validation largely supported the model, although it suggested the experts overestimated the ability of Chilean needle grass to outcompete well-managed pastures.

The model predicted that suitable habitat was more restricted than previously thought, by climate, soil, land use and land management practices (Box 3.1 Fig. 1). Nonetheless, the model indicated that eradication was not feasible. Furthermore, existing or potential barriers necessary for successful containment were absent. However, strong management action was possible, both to slow spread and minimise future impacts. Dispersal of seeds by roadside slashing was a major pathway, and one that has become critical since shire council amalgamation in 2008 resulted in slashers now being moved across much greater areas within the region. The model also indicated relatively simple management actions would limit long-term impacts on most properties, including hygiene practices, minimising disturbance caused by slashing, and maintaining healthy pastures.

Risk maps are powerful communication and decision-making tools for managers. However, as has been found more generally where science interfaces policy, they need to be fit-for-purpose, credible, and trusted if they are to be used to their full potential. The participatory approach proved valuable in generating a shared understanding of the system in the form of a rigorous and testable hypothesis. This is especially powerful when combined with quality control and independent validation of the model (van Klinken, Murray, & Smith 2015). It also

62 · Detection and Delimitation

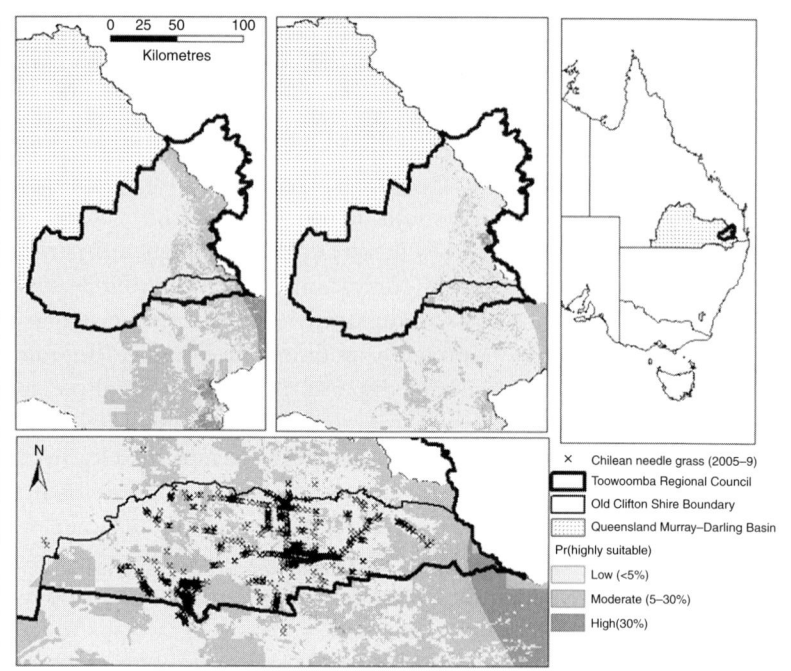

Figure 1. Area predicted to be suitable for Chilean needle grass in the south-east corner of the Queensland Murray–Darling Basin under existing (top-left panel) and optimal (top-centre panel) pasture management practices. The location of the sites in relation to the rest of Australia is shown in the top-right panel. Chilean needle grass is found primarily in the now defunct Clifton Shire (bottom panel). Roadside slashing was likely responsible for spreading plants within the Clifton Shire. Since slashing activities have been centralised following shire amalgamation in 2008, greater spread within the Toowoomba Regional Council area is anticipated.

provided decision makers with a clear understanding of the assumptions and uncertainties, and the confidence to use outcomes of the models to make sometimes tough decisions.

3.3 Surveillance Schemes

If an invasive plant is to be detected, some type of surveillance is required. Surveillance schemes can be divided into two broad groups, active and passive. **Active surveillance** is when there are specific activities designed to detect a particular invasive plant, to survey a particular area, or to

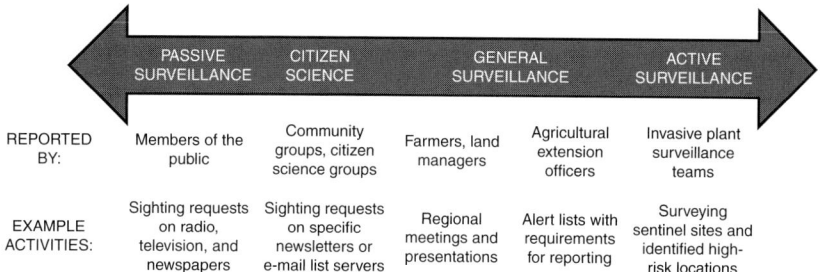

Figure 3.3. There is a continuum in types of surveillance from active to passive. Redrawn from an unpublished figure of Hester & Cacho (in review), with permission.

monitor a specific pathway. It can be resource-expensive, but detections are usually reliable, and there will be good data on sampling effort. **Passive surveillance** schemes depend on people incidentally detecting, but actively reporting, taxa. It is usually much cheaper (though can require substantial coordination) and a greater area can be covered. However, there are likely to be many false positives and sampling effort is highly variable and not necessarily proportionate to the likely sites of introduction.

In reality, of course, there is somewhat of a continuum (Fig. 3.3) and it comes down to understanding what the aim of the surveillance is. What do we want to know? How important is the reliability of the observer? What biases are there in volunteers? How quickly do we need the information? The overall trade-off is between the cost of surveillance and the time from incursion to detection. Ideally, enough effort should be placed into surveillance that a new incursion can be detected while it is at a stage where eradication or containment are still feasible options (Chapter 4).

The best examples of using active surveillance schemes to detect new incursions come from economic entomology, where pheromone traps are deployed to detect a particular species in a particular area. Plants, of course, are largely not amenable to such methods, and while pollen traps might be a useful detection method for the presence of wind-pollinated species, often it is a matter of active searching, in many cases systematically, through an area. Resources need to be put into active monitoring of harbours, border crossings, and other localities where plants are likely to be accidentally introduced due to high volumes of transport (e.g. truck stops close to borders, or wash-down areas for boats), and to likely hotspots of invasion. As such, historical plantings should be documented and periodically resurveyed for new instances of naturalisation or invasion (i.e. they can be used as sentinel sites). Similarly, it is often useful to review records of species (e.g. from herbaria and catalogues) and revisit the original collection localities. Such surveys can highlight new populations and are an essential part of establishing how widespread a species is in the country. They are not, of course, useful in detecting new introductions. Importantly, invasion hotspots are often highly degraded areas, and sometimes distinctly unpleasant places to work. Therefore, people with both generalist and specialist skills in identification need to be engaged for dedicated surveys of particular areas.

Welch, Geissler, and Latham (2014) have recently provided an excellent resource in support of active surveillance in protected areas. For guidance, this extensive report is prefaced by a decision tree, a key that poses a number of questions regarding the planning, setup, and implementation of an invasive plant early-detection programme. Key topics include setting goals and objectives, prioritising species and resources for monitoring, formulating sampling design, and managing data. It is noted that the criteria used to prioritise species for early detection (i.e. threat posed, potential range and rate of spread, feasibility of control) are identical to those used to prioritise species for control, with the exception that species that exhibit low feasibility of control actually have a high priority for detection (Brooks & Klinger 2009).

Roadside surveying is another method of surveillance that has been enormously important in determining broad-scale patterns. Surveys can incorporate both active and passive approaches. One of the most successful such schemes is the southern African Plant Invaders Atlas (Henderson 2007). This project started out as a series of dedicated roadside surveys (i.e. active surveillance), undertaken by Lesley Henderson of the Agricultural Research Council of South Africa from 1979 onwards, to document invasive plant distributions. Over time, records were increasingly submitted by other researchers and managers who were not necessarily focusing primarily on biological invasions. As such, it has become a valuable method of documenting more casual observations (i.e. passive surveillance). The southern African Plant Invaders Atlas has gradually grown to a level where the database is South Africa's primary repository of data on the location of invasive plants, and the quarterly newsletters (www.invasives.org.za/resources/sapia-news.html) are the primary method by which new incursions are reported to the invasion science community in South Africa.

The main problem with active surveillance is the cost and effort required. However, there are, of course, many eyes and ears already in the field that can potentially assist with surveillance (as in the case with the southern African Plant Invaders Atlas). Such detections are somewhat down to luck – these detections are made by people who are in an area for reasons other than searching for invasive plants – and therefore a key challenge is to increase the likelihood of any encounter being noticed and reported. Results from simulation modelling have indicated that even small increases in detection rates attributable to the involvement of the public at large can increase the probability of eradication and substantially reduce eradication costs (Cacho *et al.* 2010).

The likelihood of detection will, of course, depend on how easy it is to detect the targeted species; single plants that are large or distinctive with regard to the surrounding vegetation are more readily detected, whatever the general interests and experience of the observer. Another critical factor is the remoteness of the invasion, or to be more accurate, where it occurs with respect to potential human observers. It is to be expected that incursions that occur near population centres, or even in less densely settled agricultural regions, are more likely to be detected through passive surveillance than those where establishment is in remote regions. For example, the probability of passive detection of the bird-dispersed rainforest invader *Miconia calvescens* would be very low owing to the remote locations in the rugged terrain of north-eastern Queensland, Australia, where its invasion is unfolding, even though the species has very distinctive vegetative features (Panetta & Cacho 2012).

Nature conservation rangers are, in part, expected to perform a 'bobby on the beat' role, and so are well placed to assist with detections. But local natural historians, herbalists, farmers, and indeed anyone interacting closely and consistently with the local environment might notice, and be concerned about, changes in the vegetation. There are ways to formalise these interactions in terms of citizen science initiatives, with **spotter networks** becoming one of the most popular. Spotter schemes are where volunteers are encouraged to conduct participatory monitoring, in some cases for specific targets, and means there are more eyes out in the field looking for new incursions. Devictor, Whittaker, and Beltrame (2010) noted that the strength of citizen scientist programmes lies in the curiosity and pleasure of the volunteers in learning and observing things that they have never noticed in their most familiar places. With regard to invasive plants, these would be the farmers and producers, foresters, campers, canoeists, and naturalists, aka the citizens out in the landscape with some type of invested interest. Spotter schemes can largely be regarded as passive surveillance, in the sense that the primary purpose of people being in a particular area is usually not for surveillance. However, if spotters are given directions and training as to what to look out for and where, their activities have some similarities to active surveillance.

A slightly separate, but related, approach is identification websites (e.g. www.ispotnature.org; www.brc.ac.uk/iRecord). People submit records, which, perhaps after some discussion online, will be given a provisional identification by other users (and possibly later confirmed by an expert). The features and recognition systems are similar to spotter networks, although they are less goal-driven. Silvertown *et al.* (2015) discussed such schemes as a type of online social network which crowdsources data. Participants are ranked as contributors, with those who comment more frequently, and more reliably, receiving an increase in their status.

The rise of spotter networks has been greatly facilitated by the widespread adoption of GPS-enabled smartphones. The current trend has invasive plant programmes developing their own smartphone reporting applications and using social media to foster awareness (Newman *et al.* 2012). Citizen scientists can use their smartphones to take pictures of a potential weed, which can then be forwarded instantly to specialists for verification and included in databases. Most of these interactive apps are free to download and are very easy to use. Many provide functions that allow one to see previous sightings on a Google map. The technology engages citizen scientists in that they can see the data being used instantly and can view distribution maps in real time – they no longer need to wait for a response from a project leader. Electronic field guides and keys are readily available to help identify plants. A notable example in North America is the EDDMapS app developed by the University of Georgia (Center for Invasive Species and Ecosystem Health). First available in 2005, EDDMapS allows for real-time tracking of invasive plants linking to local and national databases.

So how has citizen science contributed to the field of invasion science and more specifically to detecting and managing invasive plants? A notable success story is that of the Invasive Plant Atlas of New England. Created in 2001, the Invasive Plant Atlas of New England is a cooperative partnership between the University of Connecticut, New England Wild Flower Society, and several federal US agencies. The Invasive Plant Atlas of New England trains volunteers in basic field identification and data management (Bois, Silander, & Mehrhoff 2011). Trained volunteers then collect data from a variety of habitats (often including areas where no invasive plants have been previously detected) and submit their records to a web-based database. These records are published on the Invasive Plant Atlas of New England website and made openly accessible to the general public, with experts verifying the identifications. Such schemes have the added benefits of raising awareness, elevating community knowledge, and can act as significant communication tools (Box 8.4).

A key challenge is to ensure that the incentive for people to participate is maintained. For example, a scheme was set up to detect fire ants (*Solenopsis invicta*) in Queensland as part of efforts to eradicate an incursion. For a short period a bounty was paid for the discovery of nests as delimitation was approached in the urban area. This arrangement could not be corrupted, since fire ant nest establishment takes some six months, although in other examples cash payments can lead to perverse behaviours. At a broad level some form of general reporting (e.g. newsletters) is also essential to maintain interest, but it is also important to have a mechanism to acknowledge and evaluate contributions, so people are aware that they are contributing to the cause and can learn how to provide better data. Such schemes, however, can be quite short-lived as they depend on ongoing interest and enthusiasm. They therefore require proper resourcing to be successful in the medium to long term.

There are, of course, particular challenges with data quality and assurance (Delaney *et al.* 2008; Newman *et al.* 2012). While people participating in such programmes tend to be more scientifically literate than the general public (Crall *et al.* 2011), citizen scientists will often need formal training to be effective. A related issue is that species might be new to a particular spotter, but hardly new at a regional or national scale. Moreover, very few people provide most of the valuable information. In essence there is a trade-off between the number of people who participate in the scheme, the value of additional participants, and the logistics of managing the process (in particular verifying records). Spotter schemes (and citizen science in general) can be seen as a solution to the limited funding and personnel needed to collect required data, but by replacing active surveillance the quality of the surveillance can become much lower than is required. All such schemes ultimately depend on correct identification.

3.4 Identification, Errors, and Lists

A key component of incursion response is to correctly identify species, but there are several challenges to this, challenges that will require dedicated resources to overcome.

Detecting a new invader often requires combining the generalist ability to notice something new in the field with the specialist ability to identify it. However, the identification training for field biologists aims at providing a broad appreciation of the most common groups in a given area, while a key component of taxonomic training is specialisation (so one can demonstrate expertise in a particular taxonomic group). For practical purposes this means specialisation in taxa common to the area where taxonomists are based, and not the species or lineages of interest for incursion response. Therefore, incursion response programmes need to encourage field biologists to report new sightings, and they must have access to taxonomists who can at least identify that a species is non-native and coordinate with international collaborators to determine the exact identity. Reporting detections of new pests to authorities for taxonomic identification is, in fact, a requirement under many plant health acts.

Such efforts are complicated by the fact that due to the invasion process the taxon might be atypical (e.g. because of hybridisation, admixture, phenotypic plasticity, or founder effects). Therefore, it is important that traditional morphological identification is supplemented by other methods (e.g. DNA barcoding and chromosome counts) (Prentis *et al.* 2008; Le Roux & Wieczorek 2009; Wilson *et al.* 2009). A major drive in this direction has been the development of DNA barcoding as a biosecurity tool. This has been extended further in the use of environmental DNA, where the presence (or at least historical presence) of an organism in a region can be determined by sampling the environment (e.g. the water in a harbour for an aquatic plant) rather than a physical specimen directly. These molecular approaches represent enticing opportunities to improve detection rates, but it is still vital to link this information to traditional morphological taxonomy. New plant incursions are likely to be spotted by the human eye first for a while to come.

Pyšek *et al.* (2013) reviewed how a lack of taxonomic expertise can impede the management of biological invasions. They concluded that incorrect identification can lead to inappropriate legislation and misinformed management decisions. For example, samples from invasive populations of crabgrass (*Digitaria violascens*) in Europe were misidentified as a rare native species *D. ischaemum*, and by the time proper identification was resolved the species had already invaded and become a dominant species in natural heathlands in the north-west of Italy, well beyond the point at which incursion response could have been considered (Verloove 2010). Another example was the taxonomic confusion between two mannagrass species in California. Low mannagrass (*Glyceria declinata*) and western mannagrass (*G. occidentalis*) had at one point been combined as western mannagrass and considered a native species. It was later shown that they are two distinct species and that low mannagrass is not native. By the time it was correctly identified, low mannagrass had invaded large areas of California and was already causing significant impacts (Gerlach et al. 2009).

The above examples of incorrect identifications of invasive plants demonstrate how incorrect identification can lead to misinformed management responses. Moreover, identification errors have been found to propagate through lists and often reappear if people consult old lists (e.g. the name *Acacia terminalis* was misapplied to *A. elata* in South Africa, and even some recent lists include both taxa). As such, lists of invasive species can have multiple potential sources of errors. McGeoch et al. (2012) identified ten errors which led to uncertainty; minimising these errors requires careful data management and clear processes for listing species and updating those lists.

One specific problem that can be encountered is that a record cannot be reconfirmed on resurvey. This might be because the initial identification was wrong, there was an error or inaccuracy in the geographical location, or the population itself is no longer present. At what point do you call it a mistake and say the species is not in the country? The main recommendation here is that the process followed needs to be recorded, the search effort clearly documented, and a note inserted into any relevant databases as to the steps followed. And when a new incursion is detected a herbarium specimen must be collected.

3.5 Detectability and Search Effort

Even if a species is known to be present in an area, individuals and populations have to be detectable for a successful incursion response; this can require a careful evaluation of detectability with specific recommendations for how monitoring should be conducted. While some invasive plants are morphologically distinct from the native vegetation, in many cases invaders might only be conspicuous for short periods, for instance when flowering (Auld & Johnson 2014), or almost completely indistinguishable (as is the case of some invasive *Phragmites* species in North America). Detectability is thus a function of both plant features and the type(s) of vegetation in which the invading species occurs. Such combinations will range from situations in which the invader represents a life form that differs from the community in which it occurs (e.g. a tree invading a grass- or shrubland) to those in which it is similar to native plants (e.g. a grass invading a grassland). In relation to the detectability of shrub and tree species within an evergreen broad-leaved forest, Chen et al. (2009) reported that differences in detectability between species mainly resulted from distinctive morphology rather than life form per se.

Brown and Noble (2005) provided first-pass estimates of the detectability (in terms of the width of a strip within which there was an 80% chance of detecting a weed, if it were present) of different weed types in various habitats. However, apart from studies that have focused on detectability of a targeted weed in a particular habitat (e.g. Moore et al. 2011a and Box 3.2), there has been no systematic attempt to produce general recommendations of how detectable general weed types are in particular habitats. The availability of accurate estimates in this regard would be extremely useful for quantifying the required search component of search-and-control efforts targeting different weeds.

While a number of studies have investigated the survey effort necessary to detect animal species if they are present, determination of the required search effort for plants is a topic that has been addressed only relatively recently. Garrard et al. (2008) approached this problem by using failure-time analysis (the operative analogy being time to detection). They were interested in the search effort that would be required to detect a rare plant species, but this work has obvious relevance to the management of invasive plants, especially when these occur at decreasing densities as a result of persistent, effective control efforts. The average time to detection was modelled as a function of explanatory variables, with the best fit to field-obtained data provided by a model that included observer experience, weather conditions, time of day, and an interaction between cover of the dominant species (a native grass) and the time of day. Interestingly, further work undertaken by others was unable to find a consistent relationship between observer experience and the probability of detection (Moore et al. 2011a). It should be noted, however, that the objective of the approach advocated by Garrard et al. (2008) is simply to determine the *presence* of a species. Determining local abundance or attempting to find all individuals during an eradication campaign is another matter.

Given the importance of detectability, this is one area where significant advances might be made in the future due to new technologies, or indeed the adoption of existing technologies. For example, dogs have been trained to detect plants as part of a management scheme for *Centaurea stoebe* in the USA, with some promising results (Goodwin, Engel, & Weaver 2010). If there were a way to harness the detection power exhibited by host-specific herbivorous insects, then the effort spent towards the end of eradication campaigns would be reduced enormously.

Remote-sensing tools have offered some potential solutions to the issues of detectability, but generally, by the time plants are large enough and/or are sufficiently dense to be detectable remotely, the incursion is so advanced that eradication is no longer feasible. Furthermore, this technology is not considered to be effective in situations where the targeted species are obscured by overstorey vegetation or where they do not exhibit either phenological differences from plants in the surrounding landscape or unique habitat associations (Young & Schrader 2014). Where containment is the management goal, remote sensing might have a place in detecting outlying populations that arise from long-distance dispersal. Much will depend on whether the signature of the species is sufficiently different from the other species with which it is associated (Bentivegna, Smeda, & Wang 2012). Advances in remote-operated vehicles (e.g. drones) might have the potential to dramatically reduce survey costs, but we suspect much more work is required for them to be cost-effective in all but very special cases for the immediate future. It would likely be counter-productive if scarce resources for monitoring were spent on such technologies at the expense of boots on the ground.

At present, in the majority of cases systematic surveys on foot are required to find all the plants in an area (Box 3.2). Hence, the effort required to detect plants comprises the most costly component of an incursion response. Populations must be visited repeatedly in order to control plants recruiting from the seed bank or previously undetected plants before they reproduce. With effective control the number of plants should decline over time, meaning that the control effort will decrease, but the search effort required generally does not; the entire area at risk must be searched during follow-up visits.

Box 3.2 *Estimating Detectability Using Search Experiments (Cindy E. Hauser & Joslin L. Moore)*

During surveys, there is always a risk that we will fail to detect every individual weed. If we understand the detection rate in our survey, then we can predict potential failures and consequences for our weed management programme. Conversely, we can design surveys that reduce risk and expected consequences to an acceptable level.

Detection rates can be estimated with careful data collection during typical weed surveys, but they can be measured more reliably by direct experimentation, especially when species are rare. In a detection experiment we simulate and replicate a range of on-ground survey conditions and measure the effort required to successfully detect weeds. We have conducted such detection experiments to support a hawkweed (*Hieracium* spp.) eradication programme in the Australian Alpine National Park (Moore *et al.* 2011a; Hauser *et al.* 2012).

The first step in planning a detection experiment is to identify influential variables. These might relate to the target weed's characteristics (e.g. size, maturity), the surrounding environment (e.g. dominant vegetation or other visual obstacles, weather), or the observer (e.g. a searcher's level of training). It can be useful to include variables that influence detection under experimental conditions but not real survey conditions, such as trampling of vegetation over repeated visits. Any potentially influential variables should be measured before or during the experiment, and controlled where possible. Traditional methods of stratifying, randomising, and replicating within the design strengthen statistical inference. However, compromises might be necessary to ensure that the experiment is practicable and captures typical surveillance behaviour.

We selected experimental sites that represented the range of environmental conditions encountered in real surveys. It is preferable to replicate each different combination of important environmental variables. For example, we composed sixteen 20 m × 20 m plots representing grass-dominated, heath-dominated, and mixed vegetation, with low and high levels of yellow-flowering plants. Plot and site size need to allow the full range of surveillance activities to be conducted, such as allowing a searcher to walk at their usual pace and gait and test their full detection distance.

Target individuals can take many forms – they might already be present at the site at appropriate densities for testing, translocated from elsewhere, or be replaced with benign mimics. In order to maximise experimental control and manage biosecurity risks, we have used a combination of propagated hawkweeds in protective plastic pots and artificial flowers and stems (Box 3.2 Fig. 1). Variation among individuals should be measured and, where possible, controlled and replicated. It is similarly desirable to control and replicate target weed densities across their natural range, randomising the placement of any introduced individuals among suitable locations.

3.5 Detectability and Search Effort · 73

Figure 1. An artificial model of a *Hieracium praealtum* plant used in detection experiments. These models were placed in the field and used to estimate detectability. The models were built by an expert in creating models for museums. Photo courtesy of Roger Cousens.

It might not be feasible or ethical to control variables pertaining to the observers, but it is often possible to measure them. It might also be possible to categorise observer types (e.g. highly trained vs inexperienced) and ensure that types are allocated evenly across survey scenarios. Human observers tend to search strategically, and it is important to avoid cues in the experiment that might alter their behaviour. While idealised sampling designs replicate target abundances in predictable patterns, this might be noticed by participating observers and used by them to increase their detection success.

Estimating detection rates often requires additional data collection beyond standard survey procedures. Site and target data can often be recorded in advance, but the time and location of detection events, and observer and target identity, can only be captured during the experiment. At a minimum the search effort and the proportion of targets detected need to be recorded. However, recording the time to detection of each individual target increases the statistical power of the experiment. In order to minimise disruption to participants, we have found that enlisting extra staff to collect these data is often beneficial (see the picture on the back cover).

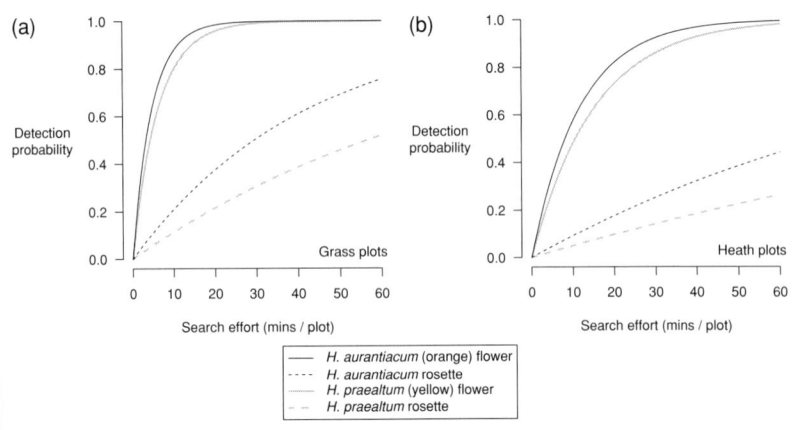

Figure 2. Typical estimates of detection probability in (a) grass plots and (b) heath plots. Based on unpublished data.

We use survival analysis to model time to detection as a function of our measured covariates. The detection time is assumed to be drawn from an exponential distribution with a rate parameter λ that depends on the observer searching, the target individual encountered, and the environmental conditions experienced via a log-link function:

$$\ln(\lambda) = a + b'x + r_O + r_T + r_E,$$

where a is the baseline transformed detectability, x is a vector of standardised measured covariates, b is the corresponding vector of regression coefficients, and r_O, r_T, and r_E are the random effect of observer, target, and environment, respectively. This model enables us to characterise the relationship between the probability of weed detection and search effort for any target–environment–observer scenario.

In our most recent experiment we found that hawkweeds are easiest to detect in grass-dominated vegetation, with detection taking longer in heath-dominated and mixed vegetation (Box 3.2 Fig. 2). Orange-flowering *H. aurantiacum* is detected most quickly, while *Hieracium* rosettes without flowers typically require extensive searching. High densities of other yellow-flowering species lead to slow detection of yellow-flowering *H. praealtum*, but it is rapidly found in the absence of this distraction.

Our detection model is now used by agencies managing hawkweeds in Victoria, Australia to help prioritise survey resources across the Alpine National Park (Hauser *et al.* 2016). Search teams can ensure that effort is sufficiently thorough to achieve their desired detection probabilities. These detection models can also inform retrospective analyses of survey data and help estimate the probability that a weed is truly absent, given that it has not been detected during a survey. Thus an understanding of weed detection rates is highly valuable for informing survey design and weed management.

Search effort is intimately related to the detectability of the targeted species at all stages of its development. The searcher might need to be almost 'on top' of plants to detect individuals, particularly seedlings or those species that are somewhat cryptic. In these cases, the search speed must be low and the pattern of search should involve close spacing of transects if an acceptable probability of detection is to be achieved. This entails a far greater search effort than would be required when searching for a plant that stands out in the associated vegetation. Intensive and frequent surveys are usually required if the problem of low detectability of seedlings is to be overcome and for seedlings not to have a chance to mature and become reproductive between searches. We return to this subject in the chapter on evaluating management performance (Chapter 5).

Timely detection and response are particularly critical for plants that occur in marine or freshwater habitats (Parkes & Panetta 2009). Invasive algae, for example, produce copious amounts of highly mobile spores – although this did not prevent the eradication of *Caulerpa taxifolia* from California. Freshwater aquatic invasive plants in riparian contexts might spread large distances very quickly, a disadvantage for management that might be at least partially offset by the reduced search area (i.e. in a linear habitat) involved, particularly if habitat requirements of the invasive plant are such that it might only persist in close proximity to water (Panetta 2012).

3.6 Delimitation

Once an invasive population has been identified, we need to know how far the population has spread. Delimitation, or determining the full extent of an incursion, is a fundamental prerequisite for a structured approach to incursion management. Once an incursion has been delimited it becomes possible to begin estimating the amount of effort (and hence investment) and regulatory action that would be necessary to achieve a particular management goal. From a biosecurity perspective, delimiting the extent of the invasion is information essential to determining whether eradication is possible. It is also important to delimit an invasion at a broad scale (i.e. how many and where are populations distributed over the landscape or region), and at a population scale (i.e. what is the extent of each population).

Delimitation can be viewed as a staged process. If the locality of the first detection is unlikely to be the same as the point of introduction, it is important to identify the putative source of introduction (i.e. trace back), then work out how the species has got from one area to the next (i.e. trace forward). This will involve consideration of potential dispersal mechanisms (hence the dispersal distances that can be expected), available introduction pathways, and suitable recipient habitats (e.g. land use types). Next, based on the current distribution, it is important to survey all likely dispersal pathways and keep going until these have been thoroughly searched. Finally, it is important to identify areas in the region with suitable habitat and to target surveys to see if there have been other introductions or additional dispersal that was not previously accounted for.

In contrast to the situation with invasive insects, where sophisticated approaches involving trapping methods have been developed, there has been relatively little research into improving methods for the delimitation of plant invasions. To assess how close a programme is to the delimitation of its target, Panetta and Lawes (2005) introduced the 'detection ratio', defined as the newly detected invaded area divided by the total new area searched, an approach that was adopted later by Gardener, Atkinson, and Rentería (2010). In the former case this ratio was used to show contrasting delimitation performances of surveys targeting *Orobanche ramosa* subsp. *mutelii* in South Australia and *Chondrilla juncea* in Western Australia. In the latter, the detection ratio was used to evaluate progress towards eradication of *Rubus niveus* on Santiago Island in the Galapagos. The major weakness of the use of the detection ratio is that its value could decline simply if the delimitation effort were poorly targeted. Any information concerning dispersal of and habitat suitability for the targeted species will therefore enhance the efficiency of the delimitation operation (Fox *et al.* 2009; Richter *et al.* 2013).

While there is as yet no published evidence of its adoption in delimitation activities, the 'Approach, Decline, Delimit' method introduced by Leung, Cacho, and Spring (2010) provided a novel theoretical approach to delimitation. This method utilises available spatial information for an invasive organism to make inferences on the possible boundaries of its invasion. For invasive plants, more populations are expected to be closer to the invasion's epicentre, with the density of plants decreasing from the epicentre to the invasion front. To follow the 'Approach, Decline, Delimit' method, the apparent boundary is neared (Approach). Next, an efficient method is designed for collecting information that will characterise the rate of decline in the proportion of invaded sites (Decline). Finally, the rate of such decline is estimated statistically to determine the likely boundary of the invasion (Delimit).

Scott and Batchelor (2014) have recently documented a delimitation exercise that was triggered by the first detection of *Chrysanthemoides monilifera* subsp. *rotundata* (bitou bush) in Western Australia. Two surveys defined the core population, covering an area of approximately 500 m in diameter, comprising 1038 plants, ranging in size from seedlings to an individual 5.5 m high that had a crown diameter of 11.6 m. In subsequent surveys, a 500 m **buffer zone** around the core distribution was searched, and five additional plants were found. The width of this zone was related to the diameter of the core distribution of the plant, being essentially an estimate of maximum effective dispersal (i.e. dispersal resulting in plant establishment) based upon the dimensions of the core. Transects were also established out to 1 km in three directions, as a basis for less intensive searching. Additional plants were found beyond the core distribution and were estimated to be either pre-reproductive or in the first or second year of flowering. Overall, a total of 246 ha was surveyed, involving 221 person hours.

Delimitation exercises can also benefit from combining active and passive surveillance. For example, at the start of the project to assess the feasibility of eradicating *Acacia stricta* from South Africa, local environmental officers and forestry managers were asked to place populations on a map from memory (no formal records had been taken up to that point). In subsequent dedicated surveys of the roads in the area, the plant was found only at the points indicated. Using correlates of the distribution, a risk map was then produced to identify areas that required regular active surveillance at landscape and local scales (these areas are now the focus of annual search-and-destroy clearing operations). As such, control efforts could be much more focused – only 17% of the road network was highly likely to contain plants (Kaplan *et al.* 2014). At a national scale, forestry plantations in areas with a suitable climate were identified, the relevant foresters questioned about whether they had seen the plant, and publicity material distributed (i.e. passive surveillance). An addendum to this story is that local forestry workers were involved in the clearing efforts. On seeing the plant, they were able to provide a few more localities that were not within the original search region (though were highly suitable when the risk-mapping exercise was extended to those areas). In general, such adaptive sampling is preferable to less flexible approaches (Maxwell *et al.* 2012).

A major issue that requires more consideration is how to determine the relative effort that should be allocated to the detection of new populations versus controlling known populations (Maxwell, Lehnhoff, & Rew 2009; Epanchin-Niell & Hastings 2010; Giljohann *et al.* 2011). In many cases surveillance and control will be similar, particularly for search-and-destroy type activities, but, as discussed by Caplat *et al.* (2014), if it is not a straightforward exercise to detect new populations, then the costs and efforts required for surveillance and control might need to be considered separately. We return to this issue in the following chapters.

3.7 Recommendations

When planning for incursion response, hotspots should be identified and monitored, since introductions and invasions tend to occur in a few locations. However, introductions can be idiosyncratic, and are ultimately due to human actions. Once an incursion is detected, an attempt should be made to trace it back to its source and trace forward to where plants might have gone; this will often require an understanding of the social history and context and not just biology. Investment in surveillance should support active and passive activities, including trained personnel regularly driving and walking through the affected areas, as well as collaborations with affected parties and the wider public. Resources must also be dedicated for proper identification. Detectability will ultimately influence the likelihood of successful control and needs to be carefully examined. Once detected, populations must be properly delimited using a variety of sampling approaches, mostly adaptive but in some cases exhaustive.

Processes	Information requirements	Deliverables
Identify incursion hotspots	• Historical patterns of detection • History of introduced species • Herbarium records of introduced species • Prediction of future pathways of spread	• Surveillance maps
Improve detection rates	• Efficacy of different methods of detection • Relevant stakeholders in areas potentially affected	• Surveillance schemes (with active and passive components as appropriate)
Delimit incursions	• Local dispersal pathways • Habitat structure • Detectability in different contexts • Likely introduction routes	• Risk map indicating where and when to search • Estimate of search effort versus efficacy

4 · Evaluation of Management Options

Key questions addressed:

- What are the first steps in an incursion response?
- When is eradication or containment of an incursion feasible?
- How should management options be compared?

A vital part of incursion response is deciding whether and how to act. There are simply not enough resources to deal with all incursions, and most incursions pose a low threat. In this chapter we focus on evaluating management options, in particular whether the goals of eradication and containment are appropriate.

4.1 Making and Documenting an Initial Decision

When an incursion is first detected, how does one decide on the appropriate management response (Table 1.1)? There are a few basic steps to take. First, a preliminary identification of the incursion is needed, in particular to determine whether it is alien or native. Second, there should be some background research as to the threat posed. Third, an initial field visit should be made to confirm the context of the incursion, including whether there are signs of naturalisation or spread.

If we have a reasonable ability to predict the likely outcome of the incursion (Chapter 2), the decision might be straightforward. For example, if there is no history of weediness elsewhere in the introduced range, there might be no immediate need to manage the incursion. If there is a mechanistic understanding of why an incursion is unlikely to either invade or cause impact, the conditions under which management is needed can be clearly stated, e.g. a species that only releases seed after fire will pose little risk of spread in an area that is unlikely to burn (Geerts *et al.* 2013). On the other hand, if there are only a few individuals of a species of known threat (e.g. a watch-list species), the plants should be controlled, or at least their reproduction stopped, until the incursion can be delimited.

In other cases, a detailed evaluation of the incursion is required, incorporating risk assessment and information from trial control work, before a goal is set.

Managers are, however, generally risk-averse, preferring to invest in the control of widespread species because the productivity gains of controlling a widespread weed are more or less certain (Finnoff *et al.* 2007). In comparison, the benefits of managing an incursion are less clear: the alien plant might ultimately have no serious impacts; the incursion response might fail; and even if all plants are removed, future incursions might be possible. Finnoff *et al.* (2007) argue that managers should adapt a risk-neutral approach, by accepting the uncertain benefits of proactive management. This approach will likely increase the overall benefits. Harris and Timmins (2009) provide another take on this issue, arguing that it is prudent to attempt to manage numerous incursions (provided that management costs for each are relatively low) in order to 'capture' the species that would ultimately cause the most damage.

Whatever the decision, it must be documented. Incursions often happen at multiple locations or at different times, and usually involve different stakeholders. The only way in which we can learn about making such decisions and improving them is if they are recorded; this is especially the case given the proven predictive value of invasiveness and weediness elsewhere, and the transferability of management knowledge around the world. Global databases (e.g. Randall 2012; van Kleunen *et al.* 2015) have provided the data for many analyses in invasion science, but they critically depend on proper documentation. Furthermore, the conditions for revisiting a decision should be clearly stated. For example, if there is a lack of socio-political will to manage an incursion that subsequently becomes a widespread invasion, the experience gained needs to inform future decisions. It is important that the lessons learned in the case of pom-pom weed are not forgotten.

A useful way of deciding which invasive plants to manage is to classify them according to both the risk that they pose and the feasibility of coordinated control. In this chapter, we focus on management programmes whose goals are eradication or containment, i.e. scenarios that lie in the bottom right corner of Fig. 4.1, where risk and control feasibility are both sufficiently high to warrant an incursion response.

As discussed in Chapter 1, management options must be considered in terms of both the intrinsic traits of a target species and the broader socio-economic and environmental context (i.e. the feasibility of coordinated control; Fig. 4.2). It might not be possible to prevent re-introduction of a

FEASIBILITY OF COORDINATED CONTROL

		Low	Medium	High
WEED RISK	**Low**	Improve general weed management	Improve general weed management & some targeted control	Monitor & protect priority sites
	Medium	Targeted control	Protect priority sites	Prevent entry & contain regionally
	High	Targeted control & protect priority sites	Prevent entry & contain regionally	Prevent entry & attempt eradication

Figure 4.1. A scheme for categorising and prioritising plant invasions on the basis of weed risk and feasibility of coordinated control. A matrix with few categories for ranking is preferable, as there is often a high degree of uncertainty in assessing invasive plants. Shaded cells denote combinations of risk and feasibility where either eradication or containment is warranted. Redrawn from the Standards Australia Guidelines, HB 204–2006 fig 8.1, with permission from SAI Global Ltd under Licence 1508-c108.

species; the plant might produce high numbers of very long-lived seeds; it might be inherently hard to detect; control measures might not be sufficiently effective; the invasive population might be invading a hard-to-reach location; or plants might be distributed across several administrative regions, making the coordination of a management response difficult. Each case presents particular, though not always insurmountable, challenges. However, perennial constraints on resource availability mean that difficult choices will need to be made. Accordingly, cost–benefit analysis is regarded as a key component of the decision-making process concerning coordinated control (Fig. 4.2; IPPC-FAO 2006). Techniques to prioritise efforts are discussed in Chapter 7.

4.2 Eradication and Containment as Management Goals

When an invasive plant is detected, eradication should be the first management option assessed, followed by containment, but, as noted earlier,

4.2 Eradication and Containment as Goals · 83

Is there likely to be sufficient political support for coordinated control?
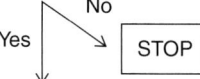

Can immigration be prevented?
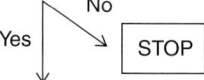

Are effective control measures likely to be available for all situations?
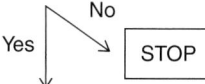

Estimate effort (resources) required to achieve eradication

Does cost–benefit analysis favour eradication over other management strategies?
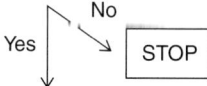

Are resources sufficient to fund the programme to its conclusion?
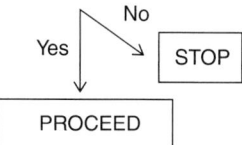

PROCEED

Figure 4.2. A decision tree for determining eradication feasibility. Note that sometimes the answer to a question will be 'maybe'. If there is uncertainty, it is worth continuing through the decision tree, as the answers to later questions can be used to inform previous ones. If STOP is reached, it is important that this decision is properly documented and the conditions under which it might be revisited clearly stated. Redrawn from Panetta and Timmins (2004) with permission.

it is dangerous to delay control until a decision is reached, and as such management will almost always need to be adaptive.

In this book we define eradication as the elimination of every single individual (including propagules) of a plant species from an area in which recolonisation is unlikely to occur (see Glossary). Extirpation, by contrast, refers to the complete removal of an individual population from an area where there is a non-negligible chance of recolonisation (or the likelihood of recolonisation has not been assessed), meaning that after extirpating a population, management resources will still be required to deal with recolonisation. Recolonisations can occur from a number of sources, e.g. from neighbouring invaded regions, from individuals in cultivation, and from new introductions to the country or region. Each potential source must be considered and managed. The success of these efforts will determine whether the on-ground management of the plant in the original area of interest is, to all practical purposes, no longer needed. After eradication has been achieved resources required for management, prevention, and awareness can be reallocated to other invasive plants, and the species can be taken off official lists of introduced taxa. As such, we see a practical as well as a theoretical distinction between eradication and extirpation.

The next management option is generally to stop or slow the spread of the invasive plant (i.e. containment). Absolute containment is achieved when there is no increase in the extent of an invasion beyond a certain distribution or line, but this is a relatively rare occurrence for invasive plant management. Containment can also be partial, often referred to as slowing spread. Both forms of containment can be economically viable propositions (Sharov & Liebhold 1998; Keller, Frang, & Lodge 2008), but slowing spread is usually the more realistic management goal (Panetta & Cacho 2012).

4.2.1 Spatial and Temporal Aspects of Coordinated Control

When setting eradication or containment as the management goal, the area and time frame considered need to be made explicit.

Knowledge of the full extent of an invasive plant incursion is a fundamental technical prerequisite for eradication (Panetta & Lawes 2005), and delimiting the spatial extent is the first step after initial detection in plant health and biosecurity programmes. Given the importance of delimitation we would recommend that targeted invasive plants should be declared *provisional* eradication targets until their spatial extent is known with sufficient confidence to provide reliable budget estimates for the effort required to achieve eradication (Panetta 2004). These estimates will also serve as inputs to a cost–benefit analysis for eradication.

A feature of many, and possibly most, invasive plant eradication programmes is an increasing awareness of how widespread the targeted species actually is as a programme unfolds. A common mistake is for the initial operational response (and budget) to focus on control, with the information available for the purpose of determining the extent of the invasion being restricted to detections made for the most part through passive (non-structured) means (Chapter 3). At least historically, a much greater investment in surveillance activities as part of plant eradication programmes has been required (Panetta 2004). It is important to note, however, that action promotes further action. If there are control operations in place which involve local stakeholders, people on the ground often feel incentivised to find and report new populations (e.g. Kaplan et al. 2014). As such, initiating trial control operations early on can often increase rates of passive surveillance.

Owing to difficulties in detecting plants when they occur at very low densities, there will always be uncertainty associated with the estimate of an incursion's spatial extent. But estimates of potential temporal durations of incursions are associated with a far greater uncertainty, particularly when persistent, hard-to-detect resting stages are involved. For example, seed persistence within some populations of invasive plants might exhibit fat-tailed distributions, such that their decline is slower than exponential (Panetta 2004), suggesting that these populations could persist at low densities for many years. Seed banks will exhibit rates of decline that vary both between and within species according to key biological characteristics (e.g. time to reproduction and seed longevity) and the ways in which these interact with control activities (Panetta & Timmins 2004). Furthermore, it is important to note that for some species vegetative dormancy is often an additional, important factor contributing to both persistence and difficulty of control (Shefferson 2009; Panetta 2015).

Eradication of an invasive plant species will not occur until seed (or vegetative) banks are exhausted, or have decreased to such low densities that the probability of extirpation in the absence of further control is relatively high. There is a tremendous range of seed persistence within plants, ranging from a matter of weeks for *Salix* and *Tamarix* species to many decades for hard-seeded species such as *Ulex europaeus* (Hill, Gourlay, & Barker 2001; Gage & Cooper 2005; Lindgren, Pearce, & Allison 2010). Clearly, the persistence of seed banks will have a major bearing on the potential time required to achieve eradication. Unfortunately, there is

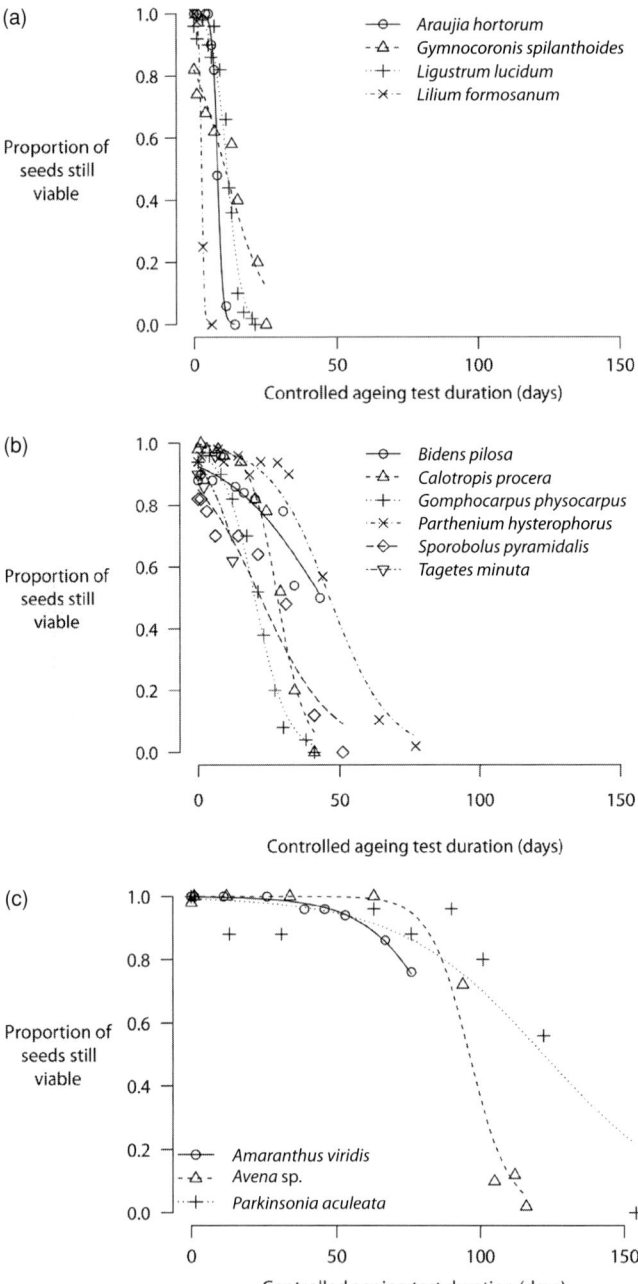

Figure 4.3. Seed survival in a controlled ageing test. Persistence of seed banks has been shown to correlate with the rates of seed survival seen under conditions of 60% relative humidity and 45 °C. Species are grouped according to whether

often little information available about this critical feature. Gaining direct estimates of seed persistence is incredibly valuable and should be a core component of research in support of coordinated control, but it can be problematic since field investigations will often require years to complete and the selection of sites (as well as the period over which sampling is undertaken) might not take into account the full range of factors that influence seed persistence.

Some studies have demonstrated correlations of seed size and morphology with seed persistence (e.g. Thompson *et al.* 2001; Moles & Westoby 2006). There is theoretical support for these correlations, given the tendency for small seeds to achieve burial readily and to avoid predation, but exceptions have been identified, so the relationship between seed physical characteristics and persistence is not robust. Recent research has demonstrated scope for using the seed survival functions observed under controlled ageing tests in the laboratory to assign invasive plant species to broad seed persistence categories: <1 year, 1–3 years, and >3 years (Long *et al.* 2008; Fig. 4.3). Such crude categories might well be adequate for the purpose of evaluation of potential targets for eradication, or at the very least provide some estimate as to how long an eradication programme might take. The use of controlled ageing methods has now been extended to a range of eradication targets (Schoeman *et al.* 2010; J. Vitelli & F.D. Panetta, unpublished data). That said, we would still recommend that wherever possible such studies are combined with field tests.

4.2.2 Extirpation and Reduction in Rates of Spread as Secondary Goals

The goals in different types of invasion management programmes are not, however, independent (Hulme 2006). For example, a programme whose overall goal is to contain an invasion might well include extirpation of outliers as part of the **management plan**, while reducing rates of spread can be an important part of an eradication programme (Panetta 2007); in

---

they exhibit: (a) transient seed banks (time to 50% viability under controlled ageing conditions (P_{50}) < 20 days); (b) short-lived seed banks (P_{50} = 20–50 days); or (c) highly persistent seed banks (P_{50} > 50 days). Data for *Parkinsonia aculeata* are for scarified seeds. Seed viability was estimated based on the germination rate of ~50 seeds that were removed from the controlled ageing test after a set period and given 30 days to germinate under conditions appropriate for the species. Redrawn from Long *et al.* (2008) with permission.

Table 4.1. *Primary and secondary goals of coordinated control programmes. The existence of primary and secondary goals reflects the fact that it is possible to eradicate an invasion even if absolute containment is not achieved (as long as populations are extirpated more quickly than they establish) and that for containment to have value it is not necessary to extirpate populations as long as rates of spread are reduced.*

Overall management goal	Primary goal	Secondary goal
Eradication	Extirpation	Reduction in rate of spread
Containment	Reduction in rate of spread	Extirpation

fact, the eradication process requires surveillance, containment, and treatment (IPPC-FAO 2006). Therefore, the relative importance of extirpation and reducing spread as goals varies depending on whether the overall goal is eradication or containment (Table 4.1).

4.3 Assessing the Feasibility of Management Goals

The feasibility of different management goals is influenced by a wide range of parameters, including biological, operational, economic, and socio-political factors (Panetta 2009), but whatever the management option, different plant incursions will be characterised by different management feasibilities. A low feasibility of eradication or containment does not mean that success is beyond reach, but without adequate investment and operational support it most certainly will be.

4.3.1 Feasibility of Eradication

The effort required to achieve eradication is a function of three elements: (1) the area which must be searched; (2) the area over which control efforts need to be applied; and (3) the difficulty of eradication in a given context (Panetta & Timmins 2004). These have been termed the **gross area**, the **net** (or **condensed**) **area**, and the **eradication impedance** respectively (though note that since gross area considers areas beyond those in which plants are actually located, it is best to always state explicitly how such areas were measured). Eradication impedance is composed of four groups of factors: accessibility, detectability, biological characteristics, and control effectiveness (Table 4.2). In the semi-quantitative model proposed by Panetta and Timmins (2004),

Table 4.2. *Factors that can impede eradication efforts. Reproduced from Panetta (2009) with permission.*

Logistic considerations	Number and spatial distribution of populations
	Accessibility of populations
Detectability	General conspicuousness within the matrix of invaded vegetation
	Detectability prior to reproduction
Biological characteristics	Reproduction through vegetative fragmentation
	Minimum length of the pre-reproductive period
	Maximum longevity of seeds or vegetative propagules
Control effectiveness	Number of treatments required to control the largest plants
	Percentage of the invaded area requiring control procedures more expensive than standard methods (e.g. proximity to watercourses might restrict use of herbicides)
	Potential for managing propagule dispersal

values for all of these factors are summed to give a value for total eradication impedance, which is then multiplied by the total gross area to give an 'eradication effort score'. In an analysis of 30 attempted invasive plant eradication programmes in Galapagos, Gardener, Atkinson, and Renteria (2010) subsequently demonstrated that values of this score (range 9–259 000) were broadly related to eradication success: the nine successful programmes all had eradication effort scores of less than 2100. Using the same approach, Buddenhagen and Tye (2015) have recently reanalysed these programmes, but did not explicitly consider threshold values for eradication effort scores.

In order to better explore the upper limits of invasive plant eradication feasibility, various theoretical approaches have been explored. Cacho *et al.* (2006) built upon the framework of Panetta and Timmins (2004) to produce an approach that emphasised species detectability. By incorporating detectability into a population model they showed that for a given level of detectability and search effort, the factors of search speed, control effectiveness, germination rate, and seed longevity had the greatest influence on eradication programme length. Monte Carlo simulations for invasive plants with different life-histories provided estimates of probabilities of eradication within various time frames.

Rejmánek and Pitcairn's (2002) work was the first formal analysis of data from invasive plant extirpation programmes (16 species and 50 populations). They found that extirpation costs increased dramatically with invaded area. No populations covering more than 1000 ha (gross area in the sense of the area that needed to be surveyed on return trips) had been extirpated and it was concluded that extirpation of populations of more than 1000 ha was unlikely, given a 'realistic amount of resources'. It is interesting to note that biological and ecological attributes of the targeted species were not considered in their analysis.

This work sparked considerable interest in the technical factors that contribute to eradication feasibility. Cunningham *et al.* (2003) surveyed experts in invasive plant eradication for their views on the probability of success for 17 hypothetical invasive plants with different life-histories, in scenarios comprising eight broad cost ranges. They found only four variables to be important: (1) the total gross area of invasion, (2) the number of populations, (3) the ease of access for control, and (4) seed longevity. Woldendorp and Bomford (2004) modelled costs (estimated from actual effort) for 20 invasive plant eradication projects in various stages of completion against a variety of attributes. Their study showed that the cost of eradication was largely determined by the total area that was treated (i.e. net area, cf. gross area in the study of Cunningham *et al.* (2003)), with no improvement to the model by inclusion of other variables. In recent work based on global data sets, Pluess *et al.* (2012a; 2012b) found that the extent of invasion has been the major determinant of the probability of eradication for a broad range of invasive organisms, including invasive plants.

Critically, there are several issues with the above studies. The first is a need for data from successful programmes that span a wide range of invaded areas (most of the documented successes have targeted very small incursions; Fig. 4.4). Second, there is an understandable bias against reporting failed programmes. Third, the data are time-censored; eradication campaigns that will take a long time do not appear in such analyses as they haven't reached completion (some useful suggestions as how to explicitly include time to eradication are provided by Dodd *et al.* (2015)). Finally, failed programmes are not very helpful when estimating the effort required for success, since it is difficult to assess under-investment as a factor independently from other causes of failure, such as poor organisation or operational performance, or a failure to overcome basic risks and constraints (Parkes 2006).

For invasive plants, scoring approaches have been applied to the entire spectrum of risk, from the assessment of risks of invasiveness and impact to post-border risk management. Findings from the field of social science

4.3 Assessing the Feasibility of Management · 91

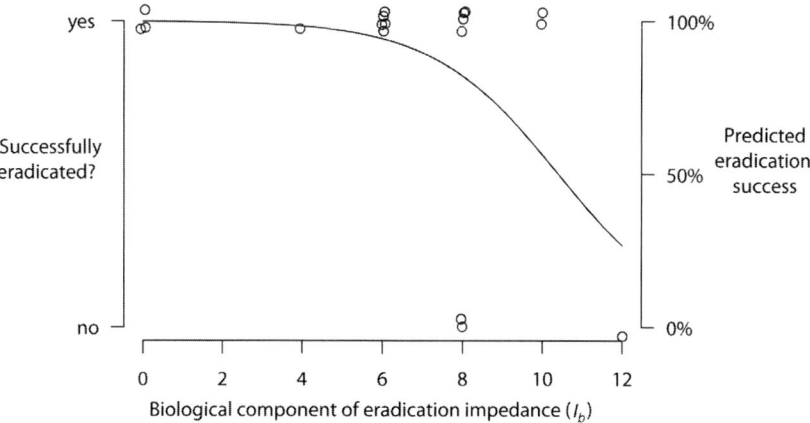

Figure 4.4. Outcome of eradication efforts in relation to the biological characteristics of the targeted species. The points on the graph are from eradication attempts (Panetta & Cacho 2014). Cases were classified as unsuccessful if either eradication was abandoned as the management goal or the targeted species reappeared after eradication was declared. Some error is added to the points to prevent overplotting. The line is a logistic regression curve fitted to the data. Species with eradication impedances greater than 10 are predicted to have less than a 50% chance of eradication, and those with eradication impedances less than six have a 95% chance of eradication. The biological characteristics (i.e. time to reproduction and seed persistence) of the targeted species are classified in terms of the eradication impedance (I_h) – see Panetta and Timmins (2004). Redrawn from Panetta and Cacho (2014), with permission.

provide strong support for the use of such semi-quantitative methods, as long as key parameters are identified and are well-coded (scored) (Kahneman 2011, p.226). Furthermore, equal weighting of parameters has been shown to be very robust for making predictions. Scoring methods have been developed and tested for eradication feasibility (Table 4.3).

Categorisation of both eradication impedance and population size as low, medium, or high yields a matrix with nine categories (Fig. 4.5). Determination of the overall eradication feasibility categories in this matrix has considerable empirical support (Panetta 2009; Gardener, Atkinson, & Rentería 2010; Howell 2012). Small invasive plant populations (gross area <50 ha) are considered to be eradicable whatever the impedance of the target species, as would be middle-sized (50–200 ha) populations of targets with low impedance. At the other extreme, large populations (>200 ha) of plants with moderate to high impedance ratings are expected to be characterised by low eradication feasibility, as are

Table 4.3. *Scoring for eradication impedance. Here we follow the approach of Panetta and Timmins (2004), but in order to simplify the scoring process, impedance factors have been reduced to those considered most critical, and as such the scores do not align with those presented in Fig. 4.4. Note also that in Fig. 4.4 only the biological component is considered, equivalent to factors (3) and (4) here. Scores are summed to give a total.*

Factor	Score	Notes
(1) How accessible are populations?	0 for easy access; 1 for moderately accessible; 2 for difficult to access	Moderately accessible incursions will include areas that pose some operational difficulty, but that might not necessarily require specialised teams (e.g. a riparian area). To be rated as difficult to access, the incursion must involve at least some sites that are very difficult to access (e.g. a ravine).
(2) Is detectability critically time-dependent?	0 for no; 2 for yes	The objective of this question is to distinguish invasive plants that are readily detectable throughout the year from those that might be detectable only for short periods (e.g. following the production of new foliage or during flowering). Species that are detectable relatively briefly provide only small windows for control prior to reproductive events that might replenish seed banks and contribute to further spread
(3) Time to reproduction	0 for >3 years; 1 for 1–3 years; 2 for <1 year	It will be more difficult to prevent reproduction of species that reproduce quickly (e.g. annuals) than those that have extended juvenile periods. Default value (i.e. if unknown) is 1.
(4) Propagule persistence	0 for <1 year; 1 for 1–5 years; 3 for >5 years	Seed persistence is one of the most important impedance factors, since it sets the minimum duration for an eradication programme. Propagules comprise either seeds or vegetative fragments (e.g. cladodes or segments for cacti). Default value is 1.
(5) Is human-mediated dispersal a major contributor to weed spread?	0 for yes; 2 for no	Weeds that are highly reliant upon human-mediated dispersal offer a relatively high potential for management of dispersal pathways. They also allow for accurate tracking of propagule movements.
Total for impedance to eradication (i.e. eradication feasibility)	≤3 is low; 4–5 is medium; >5 is high; 11 is maximum	Impedance values are assigned for an invasive plant in situations where it has been detected, not where it could be found.

4.3 Assessing the Feasibility of Management · 93

SEARCH AREA

		Low	Medium	High
IMPEDANCE	Low	9 species (e.g. *Nassella tenuissima*; *Ulex europaeus*)	1 species (*Acacia pennata*)	No species
	Medium	4 species 2 genera (e.g. *Cylidropuntia prolifera*; *Neptunia* spp.)	1 genus (*Cecropia* spp.)	5 species (e.g. *Cylidropuntia tunicata/rosea*; *Gleditsia triacanthos*)
	High	4 species (e.g. *Limnocharis flava*; *Mimosa pigra*)	6 species (e.g. *Alternanthera philoxeroides*; *Mikania micrantha*)	7 species (e.g. *Chromolaena odorata*; *Miconia calvescens*)

Figure 4.5. Categorisation of Queensland Class 1 plant incursions in relation to eradication feasibility. Impedance scores were classified as low (≤ 3), medium (4–5), or high (>5) as per Table 4.3; search areas were classified as low (<50 ha), medium (50–200 ha), or high (>200 ha) to reflect findings in the literature regarding the sizes of incursions that have been eradicated. While it is possible to eradicate incursions for which total search area exceeds 200 ha, the feasibility of eradication for these is relatively low, all other factors being equal. The numbers of taxa targeted for an incursion response are shown in each cell, with representative taxa listed in parentheses. White matrix cells represent combinations that confer high, light grey cells moderate, and dark grey cells low eradication feasibility.

middle-sized populations of species with moderate impedance ratings. This has important implications for the allocation of scarce resources when a number of species are concurrently targeted for eradication. When applied to real examples, the result can be surprising. For example, some of the current eradication targets in Queensland, Australia have very low predicted eradication feasibilities (Fig. 4.5). It is clear that this is a situation that will require difficult choices to be made if resources are not to be spread so thinly that little is achieved overall.

Recently, Panetta and Cacho's work has been extended by classifying species in terms of their dispersal characteristics, which in conjunction with the length of the juvenile stage and seed persistence, yields eight categories of eradication feasibility (Panetta 2015; see Fig. 4.6). In this work two dispersal functional groups were recognised. These corresponded to containment feasibilities that are either high (dispersal dominated by short-distance movements that occur via natural means, with or without human-mediated dispersal) or low (in which long-distance dispersal occurs through natural means and is therefore difficult to manage). The feasibility for containment is considered higher where human-mediated dispersal is prominent; even though human-mediated dispersal is capable of moving propagules very large distances, such movements are preventable and/or traceable, and thus are potentially manageable (Panetta & Cacho 2012).

4.3.2 Predicting Eradication Programme Cost and Duration

Since cost is a major consideration in assessing the overall feasibility of invasive plant eradication, several attempts have been made to predict programme duration and cost. Predicting eradication costs is particularly fraught with difficulty, owing to a tendency to underestimate these, and the substantial impact that variable costs, as opposed to fixed costs (e.g. relating to the administration of an eradication programme) have on the realised cost of an eradication campaign (Simberloff 2003; Donlan & Wilcox 2007).

The first formal theoretical attempt at predicting eradication costs for invasive plants was that of Cacho, Hester, and Spring (2007), who demonstrated that total cost would be high when low search effort was involved, but would fall rapidly with increasing search effort. The control component of the eradication operation falls rapidly as search time increases, since a more intense search effort would reduce the number of reproductive plants.

'WeedSearch', a model that provided estimates of programme duration and cost for invasive plants with different growth forms, life-histories, and occurring in different environmental settings, was developed as an extension of this work (Cacho & Pheloung 2007). This model was applied in Queensland to a list of 41 Class 1 (to be excluded, or eradicated when detected) invasive plants (Panetta *et al.* 2011c). Results indicated that all

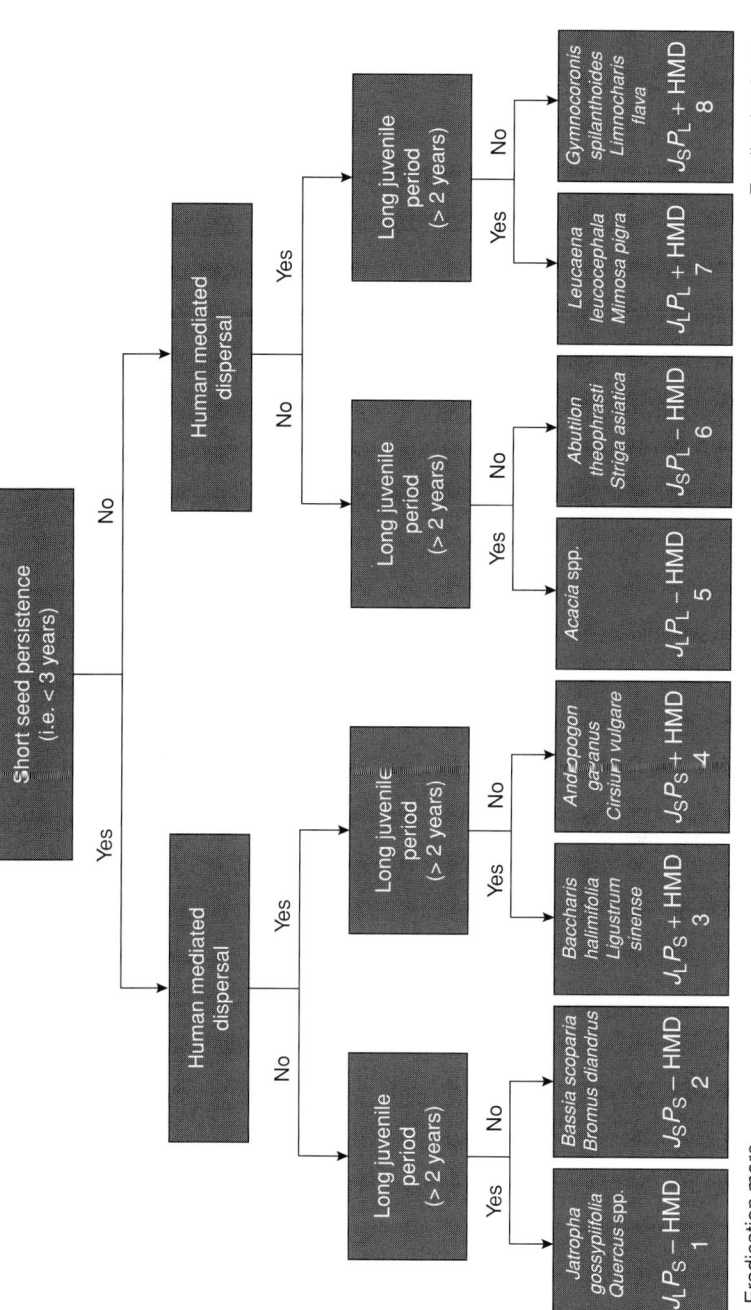

Figure 4.6. Classification of species according to an algorithm for assessing eradication feasibility. Eradication feasibility is based upon time to maturation, seed bank persistence and feasibility for containment. J_L = long juvenile period; J_S = short juvenile period; P_L = long seed persistence; P_S = short seed persistence; HMD = human mediated dispersal. Level of eradication feasibility (1–8) and examples are presented for each of the eight categories ('eradication syndromes'). Redrawn from Panetta (2015) with permission.

but one species, alligator weed (*Alternanthera philoxeroides*), could be eradicated, provided sufficient funding and labour were available, with the mean time to eradication predicted to be 17.4 years (range 2–50 years). Approximately one-quarter (i.e. $n = 10$) of the taxa could be eradicated for less than AUD 100 000 per taxon, a further 18 could be eradicated for between AUD 100 000 and AUD 1 million per taxon, while eradicating the remaining 12 was predicted to cost more than AUD 1 million per taxon.

More refined estimates of programme duration and cost, derived from a stochastic dynamic model, were presented by Panetta *et al.* (2011b). This model was based upon the rates of progression of populations from the active to the monitoring state (i.e. no plants detected for at least 12 months), rates of reversion of populations from monitoring to the active state upon further detection of the targeted species and the frequency distribution of time since last detection for all populations. It was applied to ongoing eradication programmes targeting *Orobanche ramosa* subsp. *mutelii* and *Chromolaena odorata*. The minimum periods in which eradication could potentially be achieved were 22 and 23 years respectively. Given programme performance from 2001 to 2008, cost (net present value) for the eradication of *O. ramosa* was predicted to be AUD 67.9 million (Panetta *et al.* 2011a). The minimal data requirements were a major advantage of this approach, comprising estimates of maximum seed persistence and invaded area, plus consistent annual records of the detection (or otherwise) of the invasive plant in each population. Application of this model will be considered in more detail in Chapter 5.

Empirical approaches, such as that based upon an overview of multiple eradication programmes in the Galapagos Islands (Box 4.1) can also inform guidelines relating to whether invasive plant species might be amenable to eradication or containment. Such case studies highlight the importance of habitat and environmental context in determining management feasibility. Habitat effects are accounted for in the scoring scheme in Table 4.3, but only indirectly through their influence upon the accessibility of populations and plant detectability.

Box 4.1 *Is it Feasible to Eradicate or Contain Plant Incursions in the Galapagos Islands? (Mark R. Gardener)*

The Galapagos Islands are famous for their unique biodiversity. One of the key threats to this is introduced species, including at least 871 species of vascular plants (Gardener *et al.* 2013). While 18 of these plant species are widespread and considered to be problematic for the native ecosystems, the majority were introduced recently, have yet to naturalise, and are found only in the small inhabited areas adjacent to the National Park (Trueman *et al.* 2010). Hence, an opportunity exists to employ multiple strategies to manage plant incursions, depending on their distribution, species characteristics, and the resources available.

The main strategy for the management of widespread species has been to attempt to reduce invasive plant density in areas with threatened species or of high touristic value. Over USD 3 million was spent on this strategy between 2005 and 2011, and targeted 11 species on five islands (García & Gardener 2012). For those species that were potentially invasive but with limited distributions, eradication and containment were attempted. While the goal of eradication was explicit (i.e. the removal of all individuals of a species, including seeds from a given island), containment as a goal was a fall-back position without clearly defined zonation or time frames. Since the issues affecting the success of both of these strategies are mostly shared, they haven't been differentiated for the rest of the story.

Over USD 500 000 became available between 2001 and 2007 for a pilot plant eradication programme (Gardener, Atkinson, & Rentería 2010). While many advances in plant eradication science have been made in the last ten years, the best tools available at the time were used for decision making. Candidate projects were selected using a post-border Weed Risk Assessment tool and known areas of distribution. A number of projects were not commenced because evaluation was incomplete or they were not considered feasible with the available resources. Of the 21 initiated projects, eight were successful. All of these successful projects were against targets with a distribution of less than 20 ha, on land with a single tenure, and against invasions without a well-established seed bank (Gardener, Atkinson, & Rentería 2010; Buddenhagen & Tye 2015). The remaining 13 projects were not successful owing to discontinuous funding, lack of permission from all landholders, insufficient delimitation, and tenacious biological characteristics.

Detectability and subsequent delimitation of search area was a key issue, particularly for non-tree species. The native vegetation in the humid highlands of Galapagos (where plant incursions are the most problematic) is a dense, spiny shrubland, and detecting all individuals is like finding a needle in a haystack. The three main biological characteristics that further contributed to the challenge of eradication

Evaluation of Management Options

		DISPERSAL			
		Short		Long	
		SEED BANK		SEED BANK	
		Transient	Persistent	Transient	Persistent
MATURATION PERIOD	Long	Eradication feasible (e.g. *Persea americana*)	Eradication feasible (e.g. *Acacia nilotica*)	Eradication feasible (e.g. *Casuarina equisetifolia*)	Containment feasible (e.g. *Cedrela odorata**)
	Short	Eradication feasible (e.g. *Cenchrus pilosus*)	Containment feasible (e.g. *Leucaena leucocephala*)	Containment feasible (e.g. *Aristolochia elegans*)	Neither feasible (e.g. *Rubus niveus*)

Figure 1. The three main biological characteristics constraining coordinated management, with examples from the Galapagos Island eradication programme (species from Gardener, Atkinson, and Rentería (2010) except for *).

were effective long-distance dispersal mechanisms, short maturation periods, and long-lived seed banks. Box 4.1 Fig. 1 shows the interaction between these three characteristics with respect to feasibility of eradication or containment, assuming that the target has a very limited distribution, that human-mediated dispersal could be controlled, and that all strategic/operational, biological, economic, and socio-political criteria have been met. An example of where eradication or containment was not feasible is *Rubus niveus* (hill raspberry) (Rentería et al. 2012). This species is bird-dispersed, matures in less than a year, and has a large and persistent seed bank. Since dispersal could not be prevented, the only other option was to prevent fruiting, but this was challenging because the plant matured before new individuals could be found. Hence, in three independent projects to eradicate this species on different islands, spread was not prevented despite a structured approach. Furthermore, operators could never be certain whether newly detected individuals were the result of a containment failure or had not been detected during previous surveys. Conversely, eradication of *Persea americana* (avocado) was highly feasible because its large seeds are dispersed by gravity, it is a slow-growing tree, and it has no seed bank. Alternatively, containment is a feasible option when plants have limited dispersal ability but large, long-lived seed banks (e.g. *Leucaena leucocephala*).

> In summary, eradication of plant incursions as a management goal in the Galapagos Islands is feasible only in a limited number of circumstances. Containment might be a viable alternative for some species that have amenable characteristics, but ongoing resources, collaboration with landholders, clearly defined zonation and time frames would be essential. A structured approach to monitoring and synthesis of data should be the cornerstone of all decision making and setting realistic stakeholder expectations.

4.3.3 Feasibility of Containment

A rigorous assessment of the technical feasibility of containment would need to take many factors into account (Grice 2009), most of which are also relevant to the assessment of eradication feasibility. However, containment feasibility differs from eradication feasibility in how the area under management is determined, and it focuses on spread of a plant from a site rather than its persistence there.

Since eradication is no longer the primary goal, total search area is not considered. While it is true that the greater the total invaded area, the more difficult it will be to contain a population, the relationship between invaded area and containment is complex. Strictly speaking, the total perimeter, rather than total invaded area, is the critical feature for an invasive plant incursion (Grice 2009), since the objective of containment is to prevent or reduce dispersal beyond invasion fronts. Any spread within the invaded area or **containment area** (sometimes called range filling) would be relatively inconsequential compared to spread beyond (range extension). The spatial pattern of the invasion is therefore important; many small populations would have, in total, more perimeter (boundary) than if the equivalent area existed in one or two large populations. Owing to this complexity, the feasibility of containment should be assessed primarily through factors that relate to the potential for management of dispersal pathways and the timely detection of new populations (Panetta & Cacho 2012).

All pathways should be considered when assessing the containment feasibility of targeted invasive plants. However, plants that spread primarily through human-mediated pathways appear to be particularly good candidates for containment (Panetta & Cacho 2012), in particular if the targeted species naturally spreads slowly as a result of short-distance movements, or where effective barriers either exist or can be established

(Hulme 2006). Attempts to stop or slow the spread of the targeted species are much less affected by seed bank persistence (a large component of eradication costs comes from the exhaustion of seed banks and demonstration that this has, in fact, occurred). For this reason, extirpation of outliers is not recommended as a containment strategy for invasive plants that are characterised by high eradication impedance, a major component of which is highly persistent seed banks (Panetta & Cacho 2014). However, generation time is still an important criterion since invasive plants that reproduce rapidly might spread more quickly in the absence of detection and subsequent control. The other most important considerations relate to the potential for passive (unstructured) detection and for managing propagule dispersal (Panetta & Cacho 2012). A proposed protocol for scoring containment impedance is shown in Table 4.4, though empirical work is needed to verify this approach.

The feasibility of containment also needs to be considered in relation to how core populations are managed. Where containment is the primary management goal, spread is commonly regarded in terms of movement relative to a containment line and a **barrier zone** beyond this (Sharov & Liebhold 1998; Harris & Timmins 2009; Panetta 2012). However, the maintenance of barrier zones entails costs associated with repeated search-and-control activities to ensure that boundaries are not breached. If frequent breaches occur, the establishment and maintenance of such zones will represent inefficient investment (Panetta 2012). Panetta and Cacho (2014) used two invasive plant biological characteristics (the minimum length of the juvenile stage and seed persistence), as scored according to Panetta and Timmins (2004), to construct a simple rule for when to attempt to extirpate outlier populations in an invasion comprising core and outlier populations (Fig. 4.7). This approach is supported by earlier work of Edwards and Leung (2009), demonstrating that the most important information needed to assess the feasibility of a particular species for eradication was time to maturation, because this will dictate the maximum return interval (period between site visits). No resting stage (cf. seed bank) was involved for the invasive organism they investigated (the tunicate, *Ciona intestinalis*).

4.3.4 Predicting Containment Programme Cost and Duration

Since in most cases the primary objective of a containment programme is to reduce the rate of spread of the targeted species, problems arise both in terms of demonstrating progress towards this objective (Panetta 2012) and in deciding when to cease containment efforts. As described later, one

Table 4.4. *Scoring for containment impedance. While the scoring for eradication impedance has had some empirical testing (Section 4.3.1), the scoring for containment presented here has not. It is included as an example of how we feel the field could develop, but it will likely need to be modified in response to field data. Moreover, this scheme currently only addresses the species features that relate to the feasibility of containment, but further research is required to determine the additional effects of incursion size and spatial configuration. Scores are summed to give a total.*

Factor	Score	Notes
(1) Time to reproduction	0 for >3 years; 1 for 1–3 years; 2 for <1 year	It will be more difficult to prevent reproduction (and hence dispersal) of species that reproduce quickly than those that have extended juvenile periods. Short generation times increase the chances that spread will occur undetected. Default value (i.e. if unknown) is 1.
(2) Likelihood of passive detection?	0 for high; 1 for moderate; 2 for low	Populations that generally occur either in areas of moderate-to-high human density or in other situations where it is likely that plants will be seen, will be easier to contain as it will be harder for populations to expand undetected for long.
(3) Is the species ever conspicuous?	0 for yes 2 for no	Features that make an invasive plant conspicuous at any time of the year should increase the probability of both active and passive detection. Note that an invasive plant might still be considered to be inconspicuous if it is readily confused with another common plant.
(4) Is human-mediated dispersal the major contributor to spread?	0 for yes 2 for no	Human-mediated dispersal provides opportunities for regulation of movement and for tracing pathways, both to and from a population.
Total for impedance to containment	<2 is low 2–4 is medium 5+ is high (8 is maximum)	This scoring system has not yet been tested, so these are suggested ranges only.

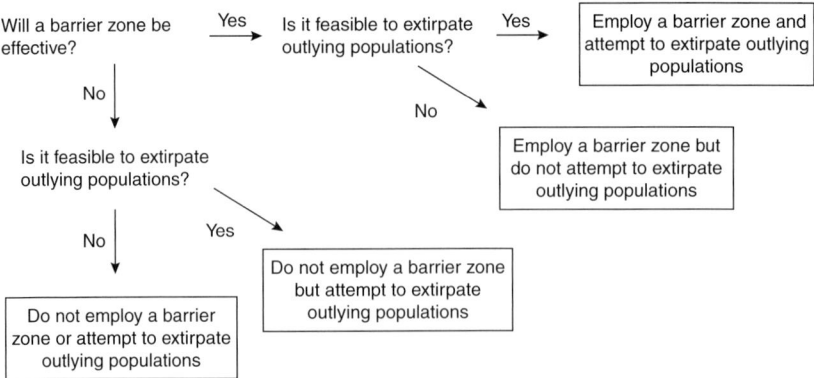

Figure 4.7. A decision tree for assigning an incursion to one of four containment strategies. Strategies differ according to whether a barrier zone is maintained around the core population and whether outlying populations are extirpated or controlled only. A barrier zone will be effective if the total amount of search effort required to achieve an acceptable target detection probability and target coverage can be supported by the available budget. The feasibility of extirpating outlying populations is based on whether the eradication impedance score based only on biological characteristics (in this case minimum time to reproduction and seed persistence) is less than or equal to a cut-off value (e.g. at $I_b = 6$ in Fig. 4.4). Redrawn from Panetta and Cacho (2014) with permission.

way to determine the endpoint to a containment effort is to use a bio-economically determined 'switching point' (also referred to as a transition point between management strategies), at which it becomes more efficient economically to control the targeted weed so that its density remains below a pre-determined damage threshold (so-called maintenance control; Simberloff 2003), rather than continuing to invest in coordinated efforts designed to reduce its rate of spread. Other alternatives are to continue containment efforts until more effective control technologies (e.g. biological control) become available, or simply to nominate a period over which the goal of containment will be pursued. Clearly an interaction might be expected between the effectiveness of containment (however measured) and the willingness to continue to adopt this strategy.

Just as the 'real' cost of eradication should be expressed as the difference between the cost of eradication and the cost of maintenance control (Simberloff 2003), the real cost of containment will correspond to the difference in effort between attempts to prevent spread (such as fecundity control near the margins of a species' distribution, or extirpation of outlying populations where this is feasible) and the effort required for maintenance control. Unfortunately, there are precious few data available on containment costs, since systematic records are rarely kept for containment programmes, as opposed to eradication programmes, where greater investment is generally made, with the concomitant need to justify expenditure on a regular basis. Ironically, one of the best documented (and effective) containment programmes was a discontinued eradication programme targeting the parasitic agricultural weed, *Orobanche ramosa* subsp. *mutelii* incursion in South Australia, which cost AUD 45 million over the 12-year period in which it was in effect. While extirpation proved impossible owing to the high persistence of the seed banks of this species, containment was near absolute owing primarily to the enforcement of a very high standard of hygiene when machinery was moved from invaded to non-invaded areas (Panetta 2012).

4.3.5 Improving Management Effectiveness

Estimates of the costs of maintenance control, eradication, and containment are, of course, based on current technologies and techniques, but there is likely room for substantial improvements in effectiveness. There have been remarkable innovations in the control of pest animals, and Simberloff (2013) has argued that, 'it seems unlikely that the biology of plants differs in characteristic ways from that of animals so as to inhibit the development of radically new control technologies'. While this remains to be proven, if general, cost-effective methods of accelerating seed bank decline could be developed, this would most certainly increase the scope for weed eradication (Chapter 9; Panetta 2015). For example, the remarkable reduction in invasive range of the parasitic agricultural weed *Striga asiatica* in the USA occurred in part because seed banks in individual populations were depleted over a period of several years with the aid of a soil-applied germination stimulant that triggered 'suicidal' germination (Tasker & Westwood 2012).

It is obviously important to determine whether current technologies are limiting containment or eradication feasibility, and to explore whether practices can be improved. Fortunately, with the assistance of the internet, practitioners can readily seek information related to the control of either the targeted species of concern or a closely related one. It should also be noted that critical biological features, such as time to maturity, might vary considerably according to location. Panetta (2015) cites a number of instances where eradication targets in the wet tropics of north-eastern Queensland, Australia, started to produce seeds much more quickly than they did elsewhere in their invasive ranges. If this information had not been determined locally, the frequency of site revisits would have been too low to prevent reproduction.

In some cases the solution might be to develop better techniques for local conditions. In others training might be required so that existing techniques are properly implemented. In still others it might be essential to devise different management structures so that the relevant weed management strategy may be implemented more effectively, in essence by improving the organisational structure (Chapter 8).

4.4 Decision Making

Decisions as to how to invest limited resources will, of course, need to be made in the context of all stages of biological invasions, and the variety of tools and approaches available to deal with them (e.g. van Wilgen *et al.* 2011). We discuss this in Chapter 7. Here we focus on determining the optimal management goal for a particular incursion (i.e. when should eradication or containment be attempted?) and there are several procedures developed to assist with this. While a number of disparate factors determine the success of an eradication or containment programme, ultimately feasibility must be viewed in the context of the amount of investment that can be made. Clearly, greater investment is justified by greater threats.

All decisions should be made in relation to an evaluation of how likely it is that a proposed programme will succeed. This assertion holds whether the primary management goal is eradication or containment. However, the success of an eradication effort is much easier to evaluate than that of a containment effort, given that the goal is relatively well-defined in the former. In the simplest terms, the effort (and therefore investment) to achieve invasive plant eradication comprises the detection effort required to delimit an invasion, plus the search-and-control effort required to prevent reproduction until extirpation occurs over the entire invaded area (Panetta 2009).

It is reasonable to expect that for any given gross area of an invasive plant invasion, the available budget will strongly influence the probability of eradication. In the end, it is up to decision makers to select the acceptable point along the curve that describes this relationship. Moore *et al.* (2011a) have demonstrated that incorporating uncertainty into feasibility analyses helps to avoid 'overly optimistic' beliefs concerning the effectiveness of management and hence the size of an invasive plant invasion that can be expected to be eradicable. They showed that if the spatial extent of an *Acacia paradoxa* invasion was unknown, attempting eradication would be cost-effective only for invasions that were less than 300 ha in gross area. In contrast, if the invasion had been delimited, an eradication attempt would be cost-effective for invasions of up to nearly 800 ha. A similar effect was apparent for whether or not to attempt containment. Other authors (e.g. Edwards & Leung 2009) have considered what probability of success would be required to constitute a worthwhile risk of investment. To do this they used cost–benefit analysis in conjunction with an eradication framework and concluded that, given the estimated feasibility of eradication for the targeted species (a sessile aquatic animal with a mobile larval phase), an eradication probability of greater than 16% would constitute a worthwhile investment risk.

A rigorous bioeconomic approach to determining the optimal management of an invasive plant incursion will need to consider invasion dynamics, the costs of control efforts and a monetary measure of damages resulting from the invasion (Epanchin-Niell & Hastings 2010). In reality, potential damages can be very difficult to estimate, especially for invasive plants whose major impacts are upon natural ecosystems. A number of studies have focused therefore on minimising the cost of eradication or containment of an incursion (Menz, Coote, & Auld 1980; Higgins, Richardson, & Cowling 2000; Taylor & Hastings 2004; Cacho *et al.* 2008). Implicit here is the assumption that the targeted species would eventually cause a very large amount of damage if it were allowed to spread without intervention. Higher control costs lead to lower optimal levels of control, which is the case with invasive plants that develop highly persistent seed banks. This will lower the net benefit of eradicating these plants (Odom *et al.* 2003; Cacho *et al.* 2008).

In the case where the net benefit of managing an invasive plant is considered, the decision to attempt eradication or containment should be based on an evaluation of the present value of benefits for the respective options over time. This approach gives rise to several switching points (Cacho 2004): the first occurs when eradication is no longer optimal for a given invaded area, but containment might be; the second occurs when the invasion is so large that coordinated management should be abandoned (Fig. 4.8). Such switching points highlight a need to have 'exit strategies' in place, whether these relate to exiting from an eradication or a containment programme. All too often, planning for eradication or containment does not include measures to be enacted if meeting either of these goals subsequently proves impossible.

106 · Evaluation of Management Options

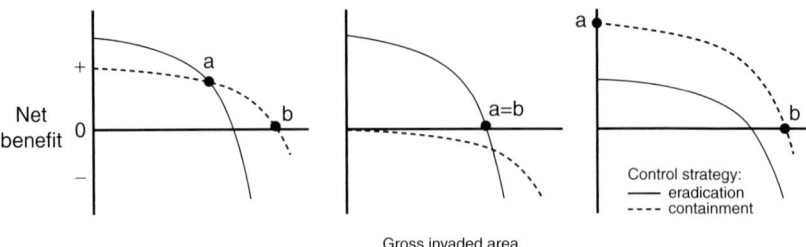

Figure 4.8. Switching points for plant incursion management strategies. The *x*-axis represents the gross area at which the plant is first detected. The solid line represents the net benefit of an eradication programme and the broken line represents the net benefit of a containment programme. In all cases, as the gross invaded area increases there is a monotonic decline in the net benefit of eradication or containment. The point marked 'a' is the switching point where eradication is no longer optimal, and the point marked 'b' is the switching point where containment is no longer optimal. In some cases the decision to attempt eradication or containment will change with the area invaded. Redrawn from Cacho *et al.* (2008) with permission.

If an invasive plant poses a sufficiently serious threat to have been targeted for eradication, there will often be justification for further investment in its containment if eradication cannot be achieved. As for the case where eradication is a primary management goal (Table 4.1), containment should be adopted as a new primary management goal only if it is considered feasible, especially since adopting this goal will involve a more sustained commitment of resources.

Screens are required to determine whether either eradication or containment is feasible for a given target. One approach to screening is through a decision tree of the type presented in Fig. 4.9. Categorisation in relation to eradication feasibility (high, moderate, and low) is made on the basis of total search area for an incursion, in combination with a measure of eradication impedance for the targeted species (Table 4.3). Species with either high or moderate eradication feasibility should be considered further as targets for eradication efforts. Where eradication feasibility is considered to be low, further categorisation in relation to containment feasibility is made on the basis of a measure of containment impedance (Table 4.4).

Programmes can be affected by unforeseen factors, including the establishment of a more serious pest organism, requiring immediate redirection of resources. This is particularly pertinent for invasive plants, which often are considered a lower priority relative to other organisms that multiply and spread much more quickly (Panetta 2009). There are

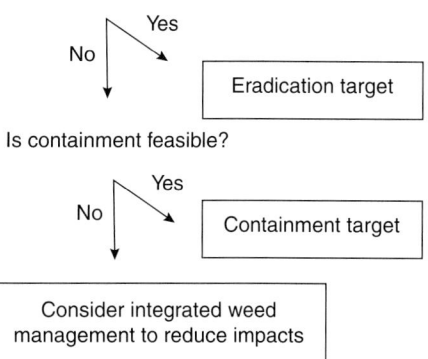

Figure 4.9. A decision tree for assigning alien plant incursions to invasion management categories. The assessment of eradication and containment feasibility can be done using the screening schemes presented in Tables 4.3 and 4.4.

obvious implications for the number (and sizes) of invasive plant eradication programmes that can be initiated and maintained contemporaneously. Agencies are constantly prioritising and re-prioritising resources between multiple established, emerging and potential targets so that the limited resources are used as efficiently as possible.

When do you transition from a goal of eradication, to a goal of containment, to a long-term management goal, to deregulation? When an agency first decides to initiate an eradication response, there also needs to be a clear strategy as to when to transition to the next management goal. This needs to be broadly communicated and articulated so that stakeholders are not confused when such transitions occur and clearly understand the reasons associated with each transition. Transparency becomes important so that no stakeholders are surprised when goals shift. However, in most cases there is no clear guidance for when to transition to the next management goal, and more importantly, agencies often fail to clearly articulate the conditions as to when eradication is no longer feasible.

When exit strategies are not identified a priori in the overall response plan, it will be very difficult to depart from a goal of eradication because proponents might continue to believe that the goal is achievable (and are fearful of losing access to funding and other forms of support, should a programme be terminated). Exiting from long-term Australian eradication programmes targeting *Chromolaena odorata* and *Orobanche ramosa* subsp. *mutelii* (Panetta 2015) was protracted and difficult, in part because endpoints had never been defined. Some guidance might be gained from programmes targeting other pest organisms. For example, the proposed goal of eradication of an incursion of the butterfly *Pieris brassicae*, first detected in May 2010 in Nelson, New Zealand, was to be reconsidered if established populations were discovered outside the Nelson operational area, if the species was clearly expanding its distribution around Nelson, or if it was still present in Nelson by the end of November 2015 (Phillips et al. 2015). However, the presence of seed banks and other factors peculiar to weeds might make extrapolation difficult. This is clearly an area that requires further investigation (Chapter 9).

If resources were not limited, the upper boundary for eradication or containment would be determined by when the net-benefit of achieving the goal was positive (Fig. 4.8). However, this is rarely (if ever) the situation, meaning that resource constraints often have a marked effect upon optimal management strategies for plant incursions. Recently, Adams and Setterfield (2015) have demonstrated how best to allocate funds to eradication and control programmes targeting the invasive tropical grass *Andropogon gayanus* under different budget scenarios. They used maximising gain and minimising loss optimisation approaches and found that under a constrained budget, allocating the annual budget to control efforts was more efficient than funding eradication efforts, owing to the current extent of invasion and apparently high rates of spread.

Given sufficient investment and the relative unimportance of other constraints, it might be possible to eradicate certain species currently designated as having low feasibility of eradication. However, the probability of achieving success against these targets will be lower than for species for which eradication feasibility is considered to be high. This is an important consideration when decisions need to be made concerning the allocation of limited resources. Any investment in the eradication of a particular incursion carries with it an opportunity cost, in that these resources are not available for managing other incursions, where the probability of success might be greater (Hester et al. 2013).

In a landmark paper that addressed the issue of whether it is more efficient to target outliers as opposed to core populations in the local eradication of *Spartina alterniflora*, Taylor and Hastings (2004) found that the optimal eradication strategy was dependent upon the budget available for control. At low to medium budgets the priority was to remove outliers, but with high budgets the optimal strategy was to target core populations first. However, given the uncertainty that often exists around budgets in the medium term (i.e. three or more years ahead), they recommended the targeting of outliers. Using a spatially explicit model for the same species, Grevstad (2005) explored optimal management, where only a fraction of the total invasion could be controlled in any one year. A strategy of prioritising outliers for control before controlling core populations achieved eradication with up to 44% less time (and effort) than where this order of attack was reversed. For invasions of *Cytisus scoparius* in natural environments, Cacho et al. (2008) found that it would be desirable to eradicate this plant from invaded areas as large as 8000 ha in the absence of a budget constraint, but once a budget constraint (typical of those faced by agencies responsible for invasive plant management) was introduced, eradication was a realistic proposition only for invasions of less than 1000 ha. Of course, theoretical studies that do not include budget constraints are often not applicable in practice.

4.5 Recommendations

Once an incursion has been detected, the various options for management need to be assessed and a decision taken as to whether to attempt eradication or containment. Feasibility should be scored in terms of the basic aspects of the biology of the species and the environmental and socio-economic context of the incursion, in particular taking the degree of stakeholder support into account. Finally, management options need to be weighed up in relation to the available resources (budget and capacity).

Processes	Information requirements	Deliverables
Initial incursion response	• Provisional identification (with formal identification to follow) • Evidence of non-native origin • Basic details of known naturalised distribution (including field notes) • A basic risk assessment (e.g. invasiveness or weediness elsewhere)	• Physical specimen in a recognised collection • Naturalisation report (if appropriate) • Documented management decision with conditions under which the decision should be revisited

(cont.)

Processes	Information requirements	Deliverables
Determining feasibilities of eradication and containment	• Basic biology (e.g. reproductive ecology …) • Value and interactions with humans • List of potential targets for management • Current distributions • Historical route and potential future pathways of introductions • Estimates of the probabilities of new introductions and of there being other as yet undetected incursions • Seed or propagule bank size • Habitat suitability or risk map • Dispersal pathways at various scales • Detectability • Management efficacy • Control and project costs • Photographs of different stages and control operations • Bioeconomic or decision support model • Cost–benefit analysis	• Priority list for management
Strategic framework for incursion response	• Estimates of likelihood of achieving a management goal under different levels of funding	• Strategies favoured under different resourcing • Decision switch points

5 · *Evaluation of Management Performance*

Key questions addressed:

- How should progress towards a management goal be measured?
- When should the management goal be changed?
- When can success be declared?

In the previous chapter we identified the factors that influence the feasibility of different management strategies targeting alien plant incursions. We also showed that extirpation and reducing spread are inter-related as goals, with the relative importance of each depending upon the overall strategic goal, whether it is eradication or containment. Of course, categorising or targeting a particular species for management and deciding whether eradication or containment options are feasible is only the first step.

The aim of this chapter is to identify and elaborate on indicators that measure progress towards the achievement of management goals. We consider how such indicators can signal impending failure, how they can be used to inform operational changes, and how they can be used by managers to decide on transitions between management goals (so-called decision switch points). Central to all of this are the data that need to be collected during the course of management programmes. The acquisition of data represents an additional cost to operational staff, and so the data need to be relatively easy to acquire. But it is absolutely critical that these data are acquired, both accurately and systematically in a manner tailored to the programme at hand (e.g. using standard forms like Fig. 5.1). Without the collection of such data, it will be very difficult to know whether progress is being made, and it will be difficult for management to be anything other than reactive.

Date: Site:

Field-work team:

GPS used:

Waypoint	latitude (decimal degrees)	longitude (decimal degrees)	Canopy width (or 2 x widths) (cm)	Height (cm)	Basal diameter (or 2 x diameters) (cm)	Number of stems	Plant Age (years)	Buds (y/n)	Flowers (y/n)	Pods (y/n)	Seeds (y/n)	Resprout (y/n)	Notes

Figure 5.1. An example of a standard field data sheet used to monitor plant populations. The exact structure will vary depending on the species involved. Not all measures will be needed (e.g. for seedlings that are clearly <1 year old, only location and approximate number need be noted, and the collecting of location data should be automated through hand-held GPS devices). Moreover, some initial research into plant allometry can provide an indication of which measures of plant size provide the best relationship with either plant age or onset of reproduction (although these can vary for resprouting individuals, with environmental conditions, and for individuals attacked by biological control agents). For cases where distinguishing individual plants is difficult, a measure of the size of clumps might be preferred.

5.1 Indicators for Management Performance at the Population Level for the Goal of Extirpation

At the population level, there need to be indicators related to both the management operations (the survey frequency and intensity in particular) and how successful the management was (plant number, location, and status in particular). These can be used to give managers some idea of whether operations are up to a standard required for an acceptable likelihood of achieving the management goal.

5.1.1 Site Survey Frequency and Intensity

The first group of indicators is related to how often and in what detail surveys are conducted.

The frequency of surveys at a site should be set such that no plants can reproduce between surveys, as preventing reproduction, while not an absolute prerequisite for eradication, is certainly desirable. Survey timing is then determined both by the **phenology** of the plant and by its detectability within the surrounding vegetation (Fig. 5.2), with surveys ideally more frequent than the minimum time to reproduction. The minimum age (or size) at reproduction can be fairly easily determined through observations at each site (or extrapolated from other sites), but detectability is much harder to estimate as it is affected by age, size, and the surrounding vegetation. Flowering will usually help detection, but in many cases it is an indication that some propagules will have been produced. Moreover, while there might be an ideal detection window, if a large number of populations is involved operational staff will find it difficult to complete search-and-control activities for all populations during the period when the target is most detectable. The frequency and timing of site visits will ultimately be constrained by the resources available, but it is important that monitoring and control activities are adequately recorded if control is to be adaptive. The recording of activities is therefore not an optional extra if resources permit, but a fundamental component of control, and must be costed as such.

The other main logistical indicator is the site survey intensity. If a plant were there, would it have been found? Site survey intensity gives an indication of: (1) the area needing resurvey; (2) the overall survey effectiveness; and (3) the performance of control teams. Survey intensity can be expressed in terms of person hours per unit area, or more spatially explicitly in terms of the distance between a searcher and any one point in

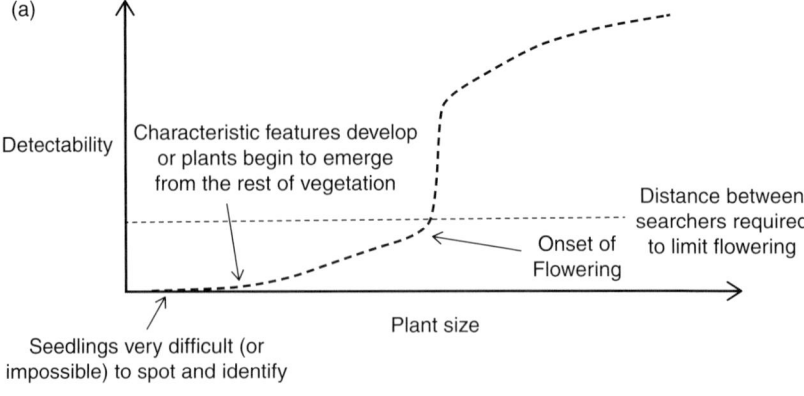

Figure 5.2. Survey intensity and timing influence the likelihood of preventing reproduction. (a) The detectability of plants is strongly influenced by their size. Detectability is quantified in terms of the distance from an observer to a plant, for a set probability of detection. For example, in order to detect at least 95% of 2 m tall plants, one might need to be no more than 10 m from any plant. This provides an indication of the required survey intensity. (b) If the objective is to prevent reproduction, and (1) seed production and flowering are continuous throughout the year or (2) there is significant overlap, then any flowering must be taken as an indication that reproduction has not been prevented, and control operations must be at least as frequent as the minimum reproductive period. However, if flowering is distinct in time from seed production (3), then there is a period when flowering can be used as an aid for detection without risking seed production.

the area demarcated for survey. Such measures are then compared against those deemed necessary in terms of the overall detectability of the plant (Chapter 3). Surveys are, of course, imperfect, and plants will be missed, but the number of undetected plants should decrease with repeated surveys, given a constant probability of detection.

Crucially, however, the probability of detection is rarely constant across a landscape. If an area is missed in one survey, it will likely also be missed on the next site visit. If there is a small dip, or a patch of deep vegetation, survey teams will tend to avoid it year on year. There are obvious biases in many surveys that affect the probability of detection, for example roadside surveys lead to sample selection bias as no data are collected past the roadside. Missing areas will mean missing plants, and repeatedly missing areas can mean missing large reproductively active plants. As such, we strongly recommend that indicators of survey performance are spatially explicit. An easy solution to this is for all control teams to carry a GPS device, and record location at suitably fine intervals (e.g. every second or metre). Of course, some plants will likely still be missed, but at least this provides a reasonable first check (Fig. 5.3). As new applications are developed, survey crews can use smartphones and tablets which have built-in GPS tracking and reporting features. Ideally there should again be a quantitative understanding of **detection thresholds** such that the uncertainty about the survey intensity can be reported, and from this a prediction made about what a given level of imperfect sampling will mean for population dynamics (and so the likelihood of achieving the management goal).

It is also important to ensure field staff remain motivated towards achieving the goal, e.g. finding plants when the density is extremely low. This might require some innovative thinking. For example, artificial models of plants can be used not only to assist with determining detection efficacy (Box 3.2), but also during training for detection, monitoring infield efficiency, and to create an incentive to search carefully. After a hard day in the field it is much more rewarding to have found something than nothing, even if it was a model rather than an actual plant.

5.1.2 Biological Indicators

The value of a given survey frequency and intensity can, of course, only be assessed by how these affect measurements of the plants themselves (i.e. number, size, status, and distribution). Such indicators assist programme managers to assess how effective surveys have been (e.g. the presence of

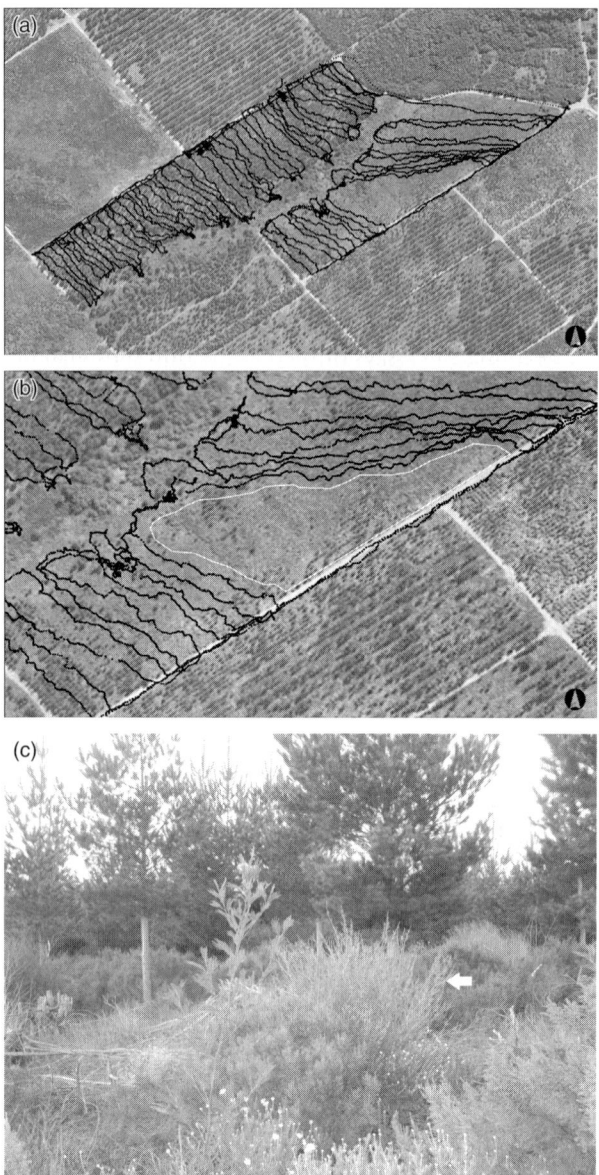

Figure 5.3. The importance of documenting surveys. The tracks shown are for two people (one of the authors (JRW) and a colleague) surveying an area as part of efforts to assess the feasibility of eradicating *Melaleuca parvistaminea* from South Africa (Jacobs *et al.* 2014). The tracks walked are shown as lines. The survey was conducted in forestry management blocks on either side of a river. While

large individuals might indicate poor performance on previous control operations) and how effective the control methods have been against the plants that *have* been detected (e.g. plants that were treated but subsequently regrew). Decisions on both the absolute and relative allocation of search and control effort can be made based on these population-level indicators (Box 5.1). Management success will be apparent when populations decrease to the stage where there are no seedlings or older plants (Panetta 2007), although the species might still be present in the form of a seed bank. For seed banks, individuals are only detectable once seeds germinate or through relatively labour-intensive measurements such as soil sampling, followed by germination assays or seed extraction plus viability assessment. It is important to note that the most informative biological indicators are often species-specific. As such, it is often useful to collect much more detailed data than necessary while evaluating eradication feasibility. These data can then be used both to inform future planning, and to determine which data should be routinely collected during future management operations in order to assess management performance most effectively (Fig. 5.1).

5.2 Indicators for Management Performance at the Programme Level for the Goal of Eradication

Should more than one population be involved, it is important to be able to assess the performance not just of managing each population, but also of the overall programme. Programme-level indicators are based upon the states of individual populations that comprise the incursion, both from an operational perspective (what percentage of sites were treated each year), and from a biological perspective (what is the current state of each population). Here we consider two methods to assess performance: the first is based on changes in the average status of all populations over time;

---

conducting the survey we forgot to watch where we were going (ironically due to a conversation about how to estimate survey intensity), and we missed a patch in an otherwise systematically surveyed block. On realising this that evening, we determined the area that needed resurveying (white line in panel (b)). The following day we returned and completed the survey. We found six plants, some >1 m in height (panel (c)), in the area that we had missed surveying the day before. Without accurate records of where we had searched we would not have realised that the area had not been completely surveyed.

and the second considers rates of progression and reversion of individual populations. Later we discuss combining indicators for delimitation and extirpation in the form of eradographs.

5.2.1 A Simple Model and Metric for Evaluating Performance towards Eradication

All populations can be classed as being in one of three states depending on the management intervention required: *active* (vegetative plants observed so active control required); *monitored* (no plants observed for at least 12 months, but area still to be regularly surveyed); and *extirpated* (no plants observed and regular surveys are no longer required) (Fig. 5.4).

Upon the detection of plants, populations previously in the monitoring state revert to the active state (Panetta 2007; Panetta et al. 2011b). Note that this classification rule is conservative; if a plant is controlled before it can reproduce, progress towards extirpation is not compromised and, somewhat paradoxically, the detection of seedlings can be viewed as a promising sign – it indicates that the soil seed bank is being depleted. Therefore, for long-lived species, and in particular species with a large seed bank and a substantial pre-reproductive period, an additional stage can be considered, where the *active* stage is split into *active (reproducing)* and *active (non-reproducing)*. If control is progressing well, then the population might be expected to be in the *active (non-reproducing)* stage for a long time and failure would only be indicated by the observation of new reproducing individuals. Shifts between the *active (non-reproducing)* and the *monitoring* categories are less critical, but extirpation can only be declared if the combined continuous period in the *active (non-reproducing)* and the *monitoring* phases is longer than the maximum seed persistence.

In theory the cut-off for shifting a state back from *active (non-reproducing)* to *active (reproducing)* is the point at which new viable seeds have been released (i.e. there is likely to be some new input into the seed bank). However, in practice the cut-off will depend on the reproductive phenology of the targeted plant. For species that have distinct phases of flowering, seed-development, and seed release (e.g. *Acacia* spp.), it might be possible to set the observation of seed pods as the cut-off, but in other cases once flowering is observed it is possible that viable seed could be present, either on detected or (more importantly) undetected plants, and so this should be the cut-off (Hester et al. 2013).

5.2 Programme-Level Indicators for Eradication · 119

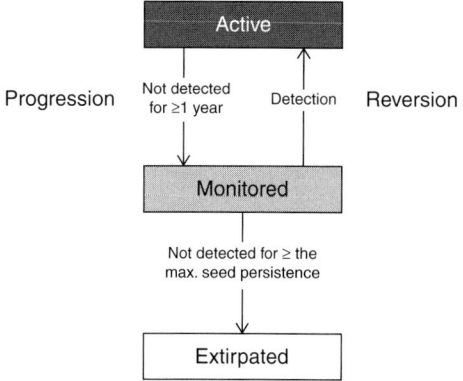

Figure 5.4. General model for classifying populations in an eradication programme. Active weed populations progress to monitored status when plants have not been detected for at least one year. Monitored populations revert to active status upon further detection of plants. A monitored population is considered to be extirpated when plants have not been detected for a period equal to or exceeding maximum seed persistence for the species. Redrawn from Panetta *et al.* (2011a) with permission.

There are two things to note with using the three-state model (Fig. 5.4). New populations cannot be accounted for until they are detected. Upon such detections, the model needs to be reset. Second, it assumes there is no dispersal back into the site, i.e. that any seedlings detected result from the seed bank. In most cases this is a reasonably safe assumption to make.

The simplest representation of this model (and so of eradication programme performance) is a collation of the status of all of the targeted populations, showing the relative proportions of active, monitored, and extirpated populations and how these proportions change over time. If the proportion of sites in the monitored category increases over time, then the programme is performing well. This was clearly not the case for an eradication programme targeting *Chromolaena odorata* in Queensland, Australia (Fig. 5.5), where almost 20 years after programme initiation few areas were plant-free (i.e. in the monitoring or extirpation state).

Building on this, an overall measure for programme performance can be developed from the proportions of populations that are in each category. The 'progress score' (P) is defined as:

$$P = \frac{e + m.w}{i}$$

120 · Evaluation of Management Performance

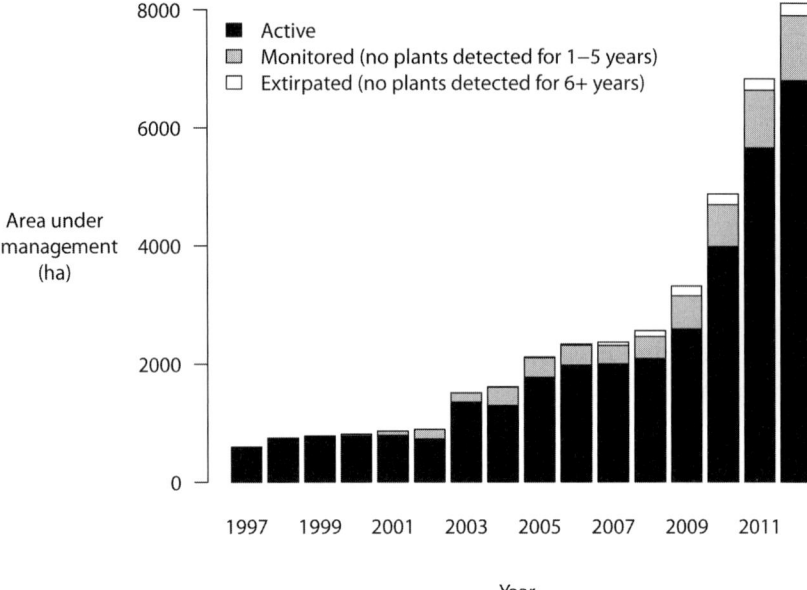

Figure 5.5. The relative proportion of the total management area in different stages for an eradication programme targeting a *Chromolaena odorata* incursion in Queensland, Australia. Almost 20 years after the start of the programme (1994), not only has the management area increased markedly, but most areas are still in the active (i.e. control) stage. Redrawn from Biosecurity Queensland (2013) with permission.

where

> e is the number of populations extirpated;
> m is the number of populations monitored;
> w is a weighting (0–1) attributed to the value of a monitored population versus an extirpated population; and
> i is the total number of populations (active, monitored, or extirpated).

The progress score P varies between 0 (plants are routinely still found at all populations such that they are all under active control and none is being monitored), and 1 (all sites have been declared extirpated, implying that the programme-level goal of eradication has been achieved). The choice of value for w should reflect the relative value of a monitored site compared to one where extirpation has been declared. This is determined by technical factors (e.g. eradication feasibility) and the significance of the extirpation milestone in terms of the potential for continued support for what is almost always a long-term management proposition. Deciding whether a population has been extirpated is basically a matter of applying a stopping rule (Section 5.5.2). Estimates of maximum seed persistence are commonly used for this purpose, although other approaches have been advocated.

It is worth noting, however, that in programmes that are succeeding, P values need not necessarily increase monotonically over time. During eradication programmes in the wet tropics of northern Queensland for *Limnocharis flava* and *Mikania micrantha*, P values occasionally declined (Brooks, Panetta, & Galway 2008; Fig. 5.6). This can be attributed to either the detection of a previously unknown invaded area or the reversion of invaded area from monitoring to active status. Both events can be expected to occur during the course of eradication programmes. As such, it is not always clear what constitutes 'adequate' or 'sufficient' progress. But if the programme is not performing well, it is best that this is apparent as early as possible, so that tactical adjustments can be made or, in the worst case, an alternative management goal can be selected (Section 5.5.2). We propose the following guidelines to interpret P values:

- Consistently low values (i.e. few or no populations in the monitoring or extirpation stage) are indicative of programme failure.
- Sustained reductions in previously moderate-to-high values are indicative of programme failure.
- If the feasibility of eradication is deemed high and progress scores are low, eradication can still be achieved if the management performance can be improved, but low progress scores and low eradication feasibility suggest that a change in management goal might be required.
- Similar to the above, higher progress scores should be required where the targeted species has long-lived propagules. Populations that revert to the active stage might replenish their seed banks and so, as far as programme duration is concerned, the time to eradication is reset.

Both of the above methods (i.e. stacked bar graphs and calculation of a simple progress score) are readily applicable, but since they do not take into account the behaviour of individual populations, very important information is lost. For example, the relative proportions of active and monitored sites might be identical in two years, but without following the status of individual populations over this period it will not be possible to determine what is going on in the monitored category. Do sites that have been in this category remain here (and thus progress towards extirpation), or have some sites reverted to active status, while others progress

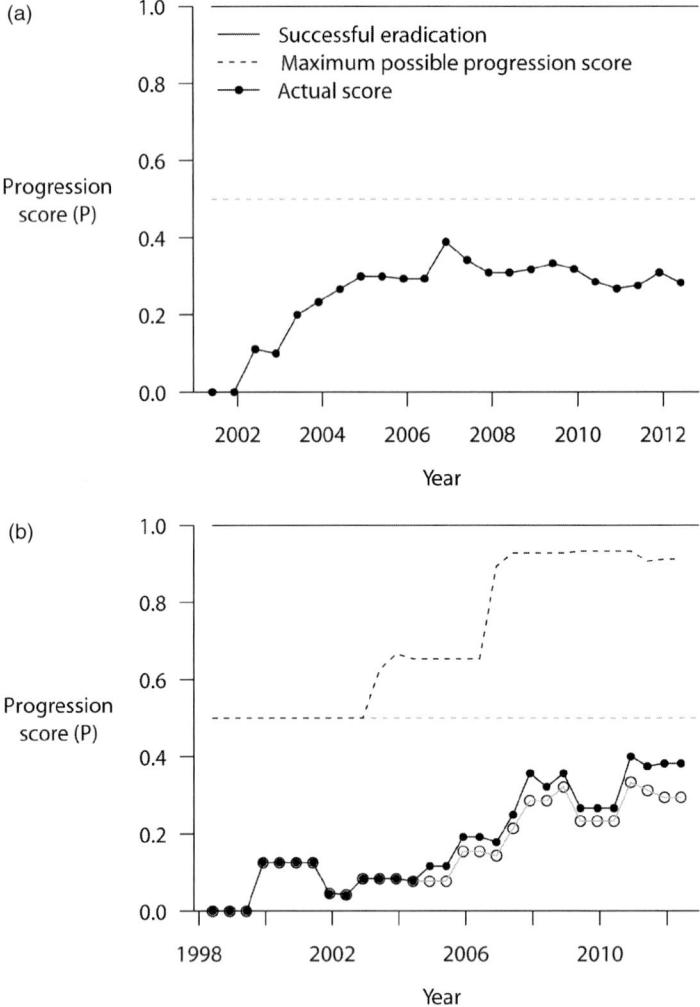

Figure 5.6. Progress score values for (a) *Limnocharis flava* and (b) *Mikania micrantha* eradication programmes in Queensland, Australia. Weighting for eradication status (*w*) is 0.5. Field studies have indicated that seed of *L. flava* is highly persistent, a proportion surviving for more than eight years (Brooks 2012). Therefore all populations are assumed to still be in the *monitoring* phase, and the maximum progress score over the 11 years shown is the weighting score (i.e. 0.5) rather than 1 (which would indicate eradication). As there is no reliable estimate of seed persistence for *M. micrantha*, two lines are presented for the programme, depending on whether populations that have been in the *monitoring* stage for longer than five years can be considered *extirpated* (solid circles, black line), or that seed persistence is greater than the programme duration, i.e. no populations are considered to be *extirpated* (open circles, grey line). Specifying the requirements to declare extirpation will affect the maximum possible progression score (cf. black and grey dotted lines on panel (b)). Data from Brooks (2012).

from active to monitored status? In the following section we show that with more thorough use of the available data, this ambiguity will be eliminated and a fairly precise picture of the performance of the eradication programme will emerge with still fairly simple metrics of performance.

5.2.2 More Complex Models for Evaluating Programme Performance

If each population is defined as active, monitored, or extirpated each year, then it is possible to calculate the rate of transition of populations between the states (Panetta *et al.* 2011b). Populations that move from the active state to the monitored state (with no detection for 12 months) can be considered 'progressions', and changes from the monitored state to the active state upon the further detection of plants are in essence 'reversions'. When a population is considered to be extirpated, its area is subtracted from the total invaded area. In order to predict the duration of an eradication programme, the model can be run either deterministically, with error terms for both the progression and reversion functions set to zero, or stochastically, via Monte Carlo simulations that sample randomly from normal distributions based on the variance of the progression coefficient and the mean square error of the linear regression of the reversion function (Panetta *et al.* 2011b).

A deterministic model based on this structure was run for the South Australian *Orobanche ramosa* subsp. *mutelii* eradication programme. This is a parasitic annual weed that invaded paddocks in dryland agricultural systems, roadsides, and, in some cases, adjacent stands of native vegetation. The status of all known populations was monitored at annual time steps. Using an estimate of maximum seed persistence of 12 years, under the best conceivable programme performance the minimum time to eradication post-2008 was predicted to be 22 years. However, when a stochastic model was used, eradication was predicted to take, on average, 62 years.

A useful extension of this is to run the model under different rates of reversion and progression to show which combination of rates is required to achieve eradication within a defined time period (Fig. 5.7). A strength of this technique is that it demonstrates graphically how programme performance can be improved – either by increasing progression rates or decreasing reversion rates. In the *O. ramosa* case this would involve either preventing seed input by uncontrolled plants (improving progression) or controlling the other plant species that function as alternate hosts (decreasing reversion).

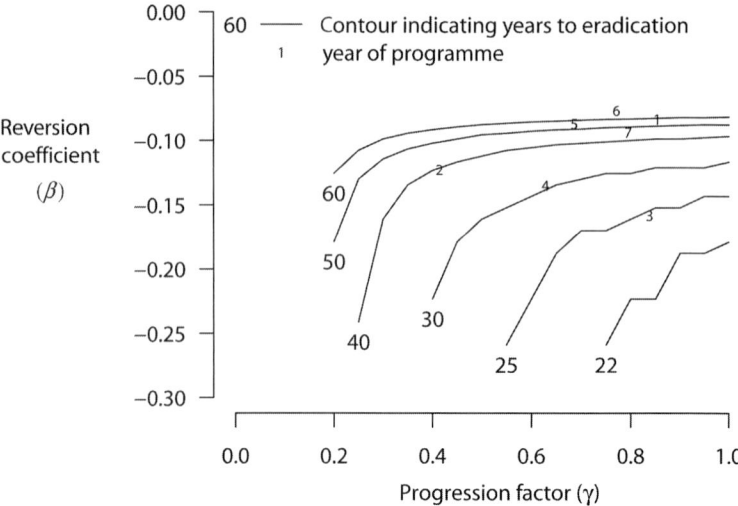

Figure 5.7. The combinations of progression factors and reversion coefficients that would allow eradication of *Orobanche ramosa* subsp. *mutelii* within specified time frames. The lines shown are isoquants, which are contour lines drawn through a set of points indicating where the same quantity of output is produced while changing the quantities of different inputs; it is a technique that is commonly used by economists to analyse trade-offs between different parameters. Each isoquant denotes the upper limit of all parameter space, allowing eradication within the respective time frame. The predicted time to achieve eradication is indicated next to each isoquant, with the minimum possible time of 22 years. The year of the programme is indicated, with year 1 representing 2002. Redrawn from Panetta *et al.* (2011a) with permission.

Using a different modelling approach, but the same observational data, Prider *et al.* (2012) estimated the proportion of fields where *O. ramosa* would be redetected at annual intervals after its initial detection. Their model incorporated detectability (including the time taken for populations to increase to a density that would permit detection), seed longevity (based upon data obtained from buried seed experiments), and what they called the 'return rate', which was defined as the proportion of previously invaded fields that were determined to be invaded in subsequent years. They predicted that after 12 years of non-detection, the probability of a single field reverting to active status was 0.01; this declined to 0.001 after 20 years. Simulation modelling revealed that the first year following initial detection of *O. ramosa* was particularly crucial for control, since fields where *O. ramosa* populations are active in this year have a large influence on the length of the eradication programme. Prider et al. (2012) predicted that eradication of the *O. ramosa* invasion could occur in 38–62 years. Note that the maximum programme duration predicted here equalled the mean programme duration from the study by Panetta et al. (2011a), indicating that the latter estimate was more conservative.

This level of sophistication in population modelling is perhaps most useful where eradication targets have highly persistent seed banks, meaning that eradication programme duration will likely extend into decades. In these cases the difference between success and failure can hinge upon improvements in practice that are prompted by insights provided by the models.

5.3 Indicators for Management Performance for the Goal of Containment

Monitoring progress towards containment is much more difficult than monitoring progress towards eradication. The endpoint of eradication (i.e. zero density) is relatively straightforward and progress towards it is easy to quantify (i.e. declines in density, then time since last detection). By contrast, measuring whether containment (either absolute (stopping spread) or partial (slowing spread)) has succeeded requires constant surveying. Failure to achieve absolute containment will be apparent and demonstrable, especially if alternative explanations such as secondary introductions or earlier failure to detect longstanding populations can be ruled out. If, however, the goal of the invasive plant management is to slow a plant's spread, a host of other considerations arises. What would the rate of spread have been in the absence of any control measures? What would the rate of spread have been if control measures focused on reducing impact rather than primarily on reducing spread? This signifies a need for estimates of what would have happened under different management scenarios: without control, with standard control operations and with specific additional operations. Other issues to be considered include the significance of spread *within* the general area over which the invasive plant is currently distributed, as opposed to spread *beyond* this area, and the time frame over which the reduction of plant spread is to be considered.

For an incursion to be contained there should be a specified containment area. This containment area is composed of a core area (that

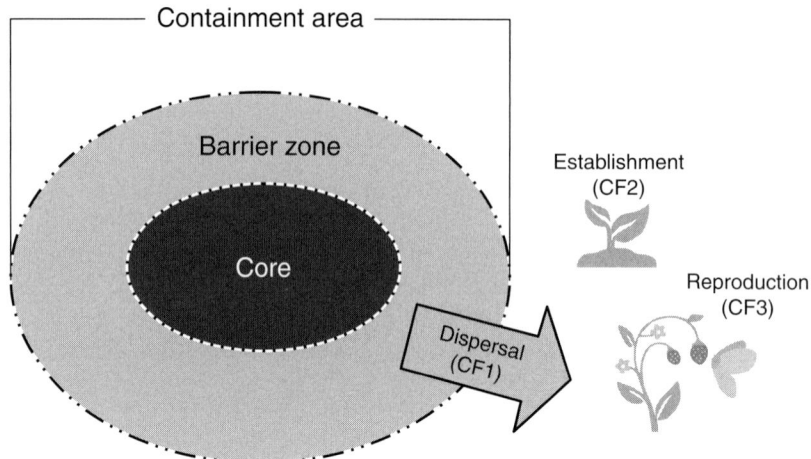

Figure 5.8. A containment area consists of a core area bounded by a population front, and a barrier zone bounded by a containment line. The barrier zone is subjected to periodic monitoring and control. Three types of containment failure (CF) are indicated (in order of severity): CF1, there is dispersal beyond the containment line; CF2, establishment is observed beyond the containment line; CF3, reproduction is observed beyond the containment line. More work is required to determine in practice where the barrier zone should be placed, and to look at the dispersal risk posed by individual plants across the landscape (e.g. Buckley *et al.* 2005). Redrawn from Panetta (2012) with permission.

is either not subjected to control measures or is part of general invasive plant management efforts), and potentially a barrier zone (where there is regular monitoring and control) (Fig. 5.8). Strictly speaking, any dispersal out of the containment area can be considered as a failure of containment. However, it is also important to determine whether this dispersal leads to establishment and/or subsequent reproduction. Specifically, it is important to know how to respond to any failures of containment. This can be by (1) extirpating any populations outside the containment area and improving containment measures implemented in the containment area; (2) increasing the size of either the barrier zone or core zone or both; or (3) abandoning containment as a management goal. Deciding the best option will require one to know how individuals reached an area outside the containment zone.

We take the approach that all newly detected populations should be considered to represent containment failure in the absence of firm evidence to the contrary (Panetta 2012). This might underestimate the

5.3 Indicators for Performance for Containment · 127

effectiveness of containment, but, because it is so difficult to prove containment failure, it would seem to be the safest approach. Next we consider the direct and indirect measures that can be used in the evaluation of containment programmes.

5.3.1 Indicators for Reducing Rates of Spread

Where human-mediated dispersal is prominent, as is the case for many invasive plant species, it can be valuable to assess the effectiveness of particular interventions designed to reduce the movement of propagules (Panetta 2012). This approach can be resolved into the monitoring of three components: (1) the *frequency* of compliance with an intervention mechanism (i.e. the proportion of vector movements in compliance); (2) the *degree* of compliance (i.e. the thoroughness of treatment for each movement in compliance); and (3) the *effectiveness* of compliance (i.e. the proportion of propagules intercepted through intervention). To the best of our knowledge, this method has not been employed, although a number of studies document contamination of machinery by seeds (e.g. McCanny & Cavers 1988; Zwaenepoel, Roovers, & Hermy 2006; Nguyen 2011). For example, in *O. ramosa*, both short- and long-distance dispersal can occur, human-mediated dispersal being largely responsible for the latter (Panetta *et al.* 2011b). Movement of machinery (e.g. tillage equipment, seeders, harvesters, tractors, and transport vehicles) appears to be the most important dispersal pathway. Strict management protocols, including wash-down procedures required before machinery can be moved from a potentially invaded area, have been critical to the containment of this plant (Panetta 2012).

5.3.2 Indicators for Success of Containment at a Programme Level

Programme-level indicators comprise the total containment area, the gross area of each population, and whether, given regular monitoring, the targeted species has been detected in the barrier zone or beyond the containment line.

Surveys for *O. ramosa* were conducted at yearly intervals within and adjacent to the containment area (Fig. 5.7), as well as on properties in other areas with links to invaded properties. When plants were detected beyond the containment area, this area was increased accordingly. Spread of *O. ramosa* can be measured in terms of changes on a number of scales: total containment area, total gross area, and number of invaded paddocks (Fig. 5.9).

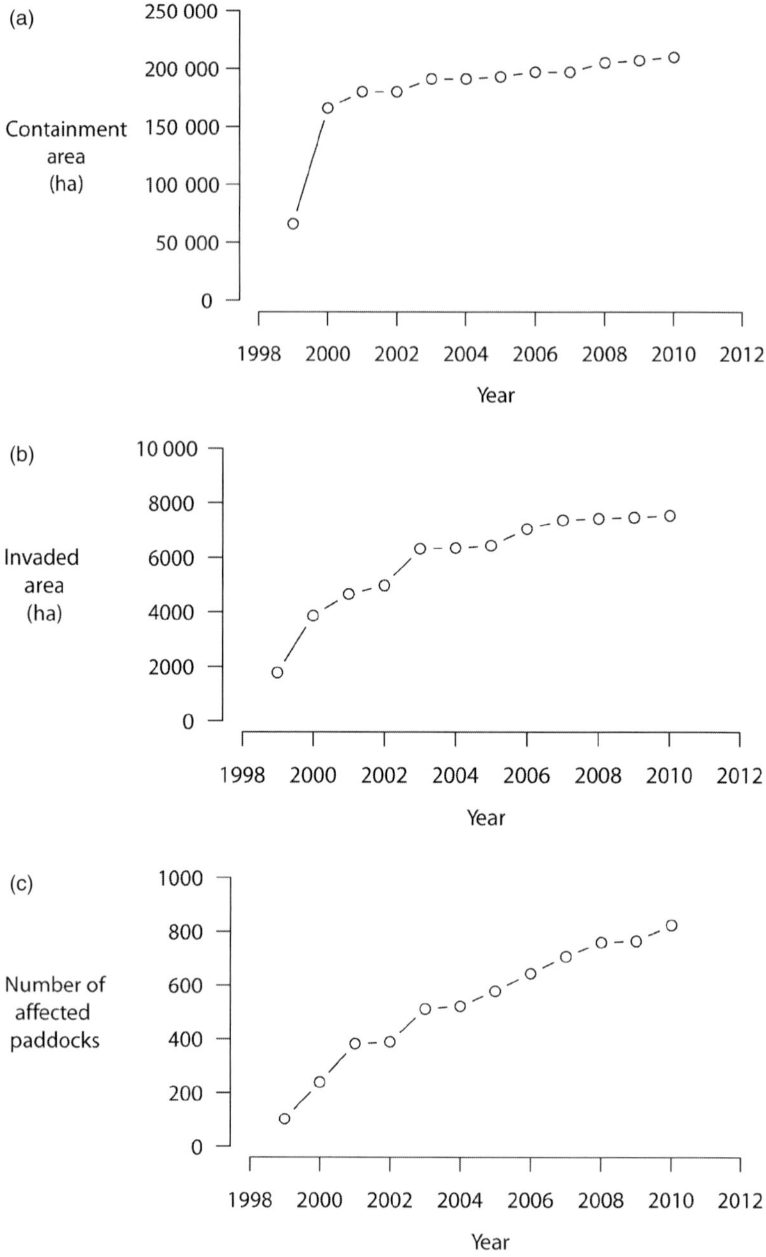

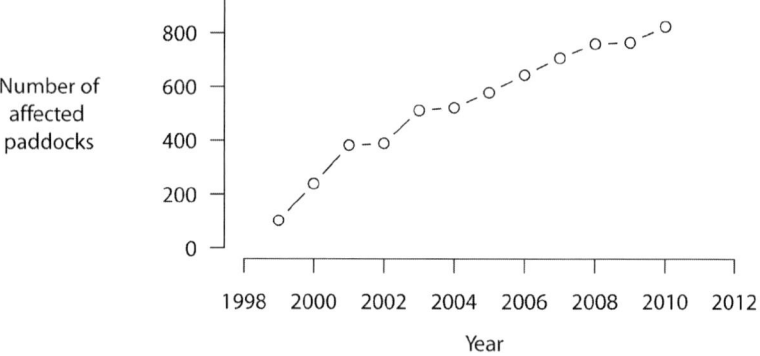

Figure 5.9. Trends in the South Australian *Orobanche ramosa* subsp. *mutelii* eradication programme indicate partial containment. The overall effectiveness of containment of this weed is evident from (a) the relative stability of the total containment area after 2000 in spite of (b) an increase in the total invaded area, and (c) a steady increase in the number of affected paddocks. The percentage of the total containment area invaded increased over time from ~2.3% to >3.5%. Redrawn from Panetta (2012) with permission.

5.4 Assessing Overall Management Performance Using Eradographs

In terms of the overall management performance it is important to consider both changes in density and extent. This has been developed in the form of eradographs, whereby measures of delimitation and progress towards eradication are plotted against the total area ever invaded (Panetta & Lawes 2007; Burgman *et al.* 2013).

The measure of delimitation is defined as:

$$D'_t = \frac{A_{d,t}}{A_{s,t}}$$

where

D'_t is the delimitation measure calculated over time interval t;
$A_{d,t}$ is the new invaded area discovered in time interval t; and
$A_{s,t}$ is the area searched in time interval t

D' will be zero under full containment, and will tend to zero as populations are properly delimited and contained over the course of a successful programme.

The measure of progress towards eradication is defined as:

$$Ex_t = E_{max} - E_{mean}$$

where

Ex_t is the measure of progress towards extirpation of all populations (i.e. eradication);
E_{max} is the time it takes to conclude that a population has been extirpated (e.g. maximum longevity of soil-stored seed); and
E_{mean} is the mean time since a plant was detected for all populations.

Successful programmes will see Ex decline to zero (though note negative values are possible).

Using field-based data from the programme targeting *O. ramosa*, the eradograph highlighted good performance against the delimitation objective but very little evidence of successful extirpation (Fig. 5.10). The latter situation was basically unavoidable, since seed banks of *O. ramosa* are highly persistent ($\geq$ 12 years). Further work (Panetta & Cacho 2014) has shown that *O. ramosa* was not a good target for eradication, despite being an excellent target for containment (Panetta 2012).

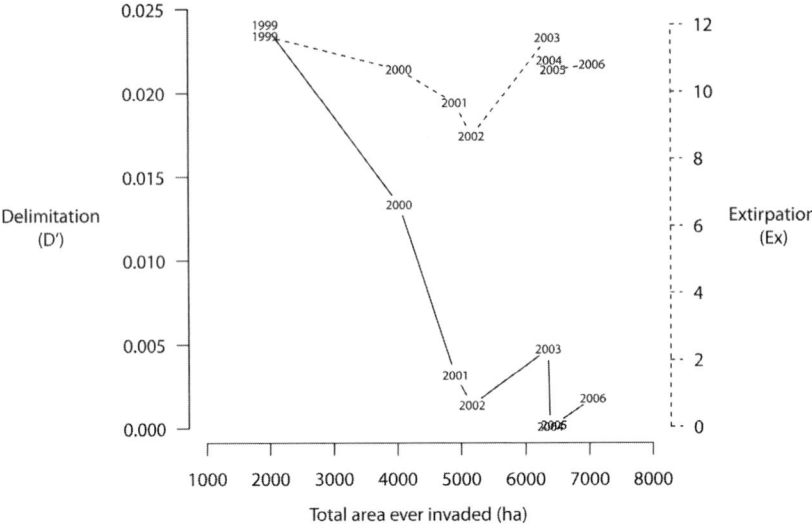

Figure 5.10. Eradograph for the *Orobanche ramosa* subsp. *mutelii* eradication programme. The eradograph combines two metrics. Delimitation (D') is shown as a solid line. It is the invaded area newly detected relative to the total area searched in a given time period. As such, D' will be zero under full containment, and will tend to zero as the invaded area is delimited and the invasion is contained over the course of a successful programme. *Extirpation* (Ex) is shown as a dashed line. It is calculated as the time it takes to conclude that a population has been extirpated minus the mean time since a plant was detected for all populations; successful programmes will see Ex decline to zero (or negative values). In the programme shown in the eradograph, extirpation was not achieved in any of the hundreds of populations comprising the invasion, even though the invasion was well delimited and contained (Fig. 5.9). Redrawn from Burgman et al. (2013) with permission.

There are various other papers that have used this, or a different formulation. For example, the spatial component of the eradograph approach was subsequently used by Brooks et al. (2009) for eradication programmes targeting *Clidemia hirta*, *Miconia nervosa*, and *M. racemosa* in far north Queensland, Australia, and by Rentería et al. (2012) for a programme targeting *Rubus niveus* in the Galapagos Islands.

There are various possible ways to extend this analysis. For example, for the delimitation metric one can take into account differences in detectability between sites and the differences in probability of occurrence at any one site, and it can be more informative to calculate the extirpation metric separately for each site (not least because in the above formulation the extirpation metric can be negative even if eradication hasn't been achieved, e.g. if most sites have been surveyed longer than necessary but individuals persist at one site). It is also likely that as a programme proceeds there will be less uncertainty in the time required to declare extirpation at a site, and more refined estimates of which sites are likely to become invaded in the future.

The purpose of the eradograph and the other evaluation methods described above is to provide decision makers with information pertinent to whether to maintain or change the overall management goal. Such information can also be employed to determine the best allocation of resources between programme activities, a topic that is addressed next.

5.5 Decision-Making During an Ongoing Management Programme

Because invasive plant management programmes are generally very protracted undertakings, it is essential that they are reviewed regularly to ensure that programmes are on track to meet their objectives within current and projected budgets (Panetta 2009). Given competing demands, decision makers will likely be mindful of the possibility that perennially scarce resources could be invested more effectively elsewhere. In particular, during the initial stages of an incursion response more resources are typically allocated towards control than to the surveillance activities necessary to provide a reliable estimate of total extent (Section 3.6). Box 5.1 addresses the allocation problem, illustrating the point that the optimal allocation of budget will change as the invasion response evolves.

> Box 5.1 *Allocating Resources (Oscar Cacho)*
>
> The decision problem faced by invasive plant managers is illustrated in Box 5.1 Fig. 1. The general principles are the same whether the management goal is eradication or containment: an agency has a limited budget to control an invasion and must decide how to allocate it to alternative activities. The goal should be to use the budget as efficiently as possible. In order to design an appropriate strategy the agency needs to determine the extent of the invasion (delimitation), while at the same time deciding whether to monitor and treat known populations. The incursion response strategy regulates the allocation of resources based on the best information available, here represented as a probability map. Part of the budget is allocated to research, to learn

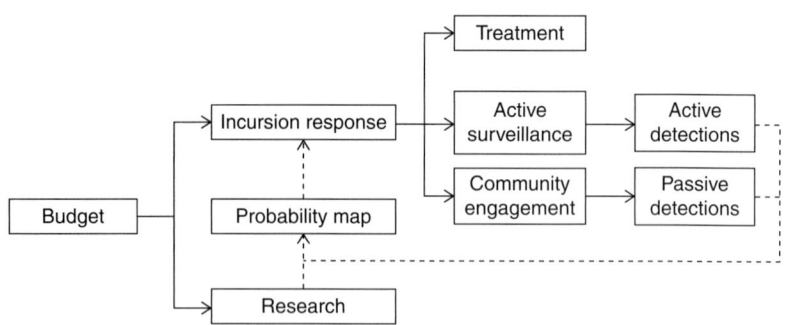

Figure 1. The budget allocation problem for control of invasive plants.

more about the invasion and its extent. This research helps improve the probability map. (Note that the 'probability map' might not be a detailed map containing actual probabilities, but could be as simple as a priority list of sites to be monitored and treated as necessary.)

The portion of the budget available for incursion response will go towards treating known populations, surveying sites of likely plant presence, and community engagement to encourage the public to keep an eye out for the plant and to report populations. Active detections and passive detections both contribute to improvements in the probability map.

The optimal allocation of the budget among the activities illustrated in Box 5.1 Fig. 1 will change as the incursion response evolves. In the early stages, when little is known about the invasion, more funds should go towards research to improve the probability map. Once the programme is well established and the probability map is reliable, there is less need for research and a larger proportion of the budget should be allocated to surveillance and treatment (Baxter & Possingham 2011).

The optimal allocation of the budget between active surveillance and community engagement would depend partly on the gross area invaded and partly on the features of the plant and the environment invaded (Cacho *et al.* 2010). If the gross area invaded is small relative to the budget, active surveillance would be able to cover the full area and no community engagement would be necessary. However, if the gross area is large relative to the budget, active surveillance by trained personnel will be unable to cover the full area, and help from the public would be required, especially to detect outliers. For certain plants and environments passive detections would be rare, such as where the plant is difficult to distinguish from native species or for inaccessible sites, or

5.5 Decision-Making in a Management Programme · 133

where the human population density is low and therefore the probability of someone encountering the plant is also low. In these cases investment in community engagement would be wasteful. Cacho and Hester (2011) show how allocation between passive and active surveillance can be assessed based on eradication probabilities and costs.

The optimal allocation to treatment depends on the stage of the invasion, the search mode and the types of control options available. In many cases active surveillance and treatment occur simultaneously, such as the case of ground search by control teams spraying or pulling the weeds they find. However, treatment can occur in the absence of active surveillance when broad-scale methods, such as aerial spraying, are available to cover the gross area. Biological control is another form of broad-scale treatment that does not require weeds to be detected through surveillance. The use of this method might become the optimal strategy when eradication is abandoned as the management goal.

A typical case of surveillance and treatment occurring separately is remote sensing, in which digital images are analysed using computer algorithms. Sites classified as invaded by this automated method must be visited in order to be treated. The allocation problem is simple if the budget is large enough to suppress reproduction by visiting all sites identified as invaded and treating them as necessary. Some of the sites classified as invaded will be false positives and visiting them will incur a cost while not reducing the weed population. This is not a complete waste because visits that confirm absence of the weed help to improve the probability map and might also help to improve the digital search algorithms. However, it is desirable to keep false positives to a minimum, especially when the budget cannot cover immediate treatment of all suspected sites to prevent reproductive escape (Spring & Cacho 2015). In this case sites must be prioritised based on the risk of escape and damage as well as on the probability that a site is a true positive. Hauser and McCarthy (2009) address the problem of spatial allocation of search effort in a heterogeneous environment and with a budget constraint.

In terms of temporal allocation, there is a trade-off between total cost and duration of an eradication programme (Hester *et al.* 2010; 2013), illustrated in Box 5.1 Fig. 2. The negative relationship arises because high surveillance effort early in the programme results in shorter times to eradication and lower uncertainty in programme duration, but at a higher cost. In deciding on the budget for incursion

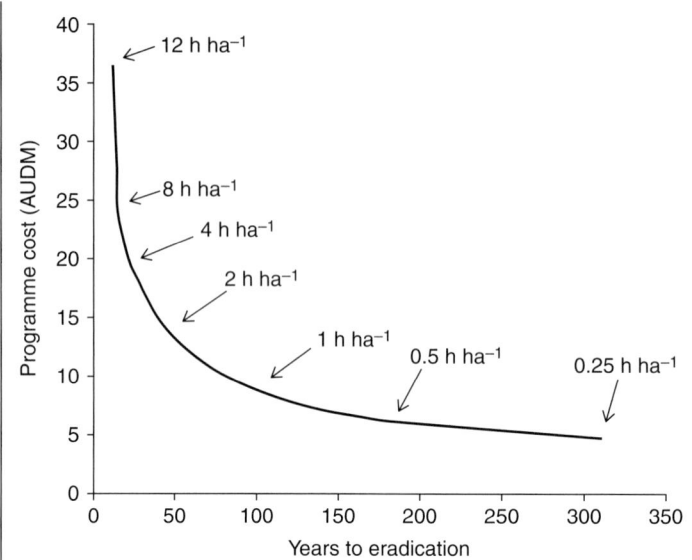

Figure 2. The trade-off between expected cost and time required to eradicate an invasion, based on search intensity (hours per hectare) for a fixed invaded area. Redrawn from Hester *et al.* (2013) with permission.

response, at least three factors need to be considered: (1) the opportunity cost of these funds not being allocated to other projects; (2) the damages associated with the presence of the weed for a longer period; and (3) the risk that the weed will escape from the managed area (Hester *et al.* 2013). This latter factor is of particular concern because every year there is a non-trivial probability that the weed will escape; the longer the programme duration, the more likely this is to happen at least once.

To capture this risk, we need to consider the future costs of escape, which also come into the decision of when to declare an area free of weeds (Regan *et al.* 2006; Rout, Hauser, & Possingham 2009; Rout, Salomon, & McCarthy 2009).

5.5.1 When to Change Goals

There is a need for certainty with regard to the availability of resources for planning and sustaining capability and expertise, as well as ongoing political support. So while the solution of changing goals might be easy on paper, such flexibility is often much more difficult to achieve. It is therefore important that the possibility and conditions under which goals can change are clearly thought through and articulated.

The optimum management of invasions depends on the expected costs of control relative to the available budget (Section 4.4). The net benefits of managing the spread of a weed population are critical to this analysis, since the decision to eradicate or contain is based on evaluating the present value of net benefits of each option over time (Cacho 2004). As discussed in Section 4.4, there are two switching points at which the control decision changes, based on the size of the invasion when first discovered. The same rationale is relevant as an eradication or containment programme evolves; if the weed continues to spread or if discoveries of significant invaded area occur, it might be prudent to change the overall management goal or to abandon any sort of species-specific coordinated control and resort to maintenance control (Simberloff 2003). Given that the goal of most containment programmes will be to slow (rather than halt) spread, it is likely that this decision will be based on the proportion of the potential range invaded and the cost of a continuation of efforts towards containment, in the context of competing demands for limited resources and political considerations. National plant protection organisations also transition between management goals as the spatial distribution changes and will move to deregulate a pest if it no longer meets the IPPC definition of a quarantine pest.

Recent work by Fletcher *et al.* (2015) has shown that resorting to containment is not always the best option if an eradication programme is not performing well. They undertook a generalised analysis of the costs of eradication and containment that would be applicable to any plant invasion for which population size, dispersal distance, seed bank persistence, and the economic discount rate are known or specified. This analysis showed that containment will cost less than eradication only when the size of the occupied zone exceeds a multiple of the dispersal distance determined by seed bank persistence and the discount rate. Furthermore, containment becomes proportionally cheaper than eradication for invaders with smaller dispersal distances, more persistent seed banks, or where larger discount rates are employed.

Changing goals should not necessarily be seen as a failure. In the absence of perfect information about what will happen, the optimal solution is to invest resources in the strategy that will perform best given the relative likelihood of different outcomes. Programmes that start off with efforts aimed at eradication will likely have had an impact on the rates of spread (though at a higher cost), and so in many cases it is best to start off with an aim of eradication. However, extreme events, such as major storms or floods, might prove 'game changers' through their effects upon propagule dispersal. A good example of this is provided by the management of *Parthenium hysterophorus* in Australia, where attempts have been made to contain its invasion over a large scale (Panetta 2012). This invasive plant is usually highly amenable to containment since its dispersal is largely human-mediated. Its seeds are usually dispersed only short distances by natural vectors such as wind, but major floods can disperse seeds over very large areas, thereby severely compromising containment efforts.

A more general challenge is to identify when it is worthwhile to invest in learning or research to reduce uncertainty, given that research will take resources away from the on-ground eradication or control efforts (Baxter & Possingham 2011). One method of analysing this is to look at the value of perfect information, i.e. if one knew what the actual state was, would it change the management decision (Moore *et al.* 2011b)? The general conclusion from this is that basic research and survey work are always needed to get a coarse estimate of the state of a system, and that intensive additional research is particularly valuable if the system is close to a decision switch point.

5.5.2 Declaring Eradication

Given excellent progress towards eradication, another equally challenging question is when to consider a plant incursion to be eradicated. This is a vexed issue, owing both to the difficulty of detecting plants at very low densities and the uncertainty relating to maximum seed persistence.

In practice, such stopping rules have been somewhat arbitrary – extirpation of a population considered to have been achieved when no detections of the targeted plant have been made for two or three years. Stopping rules for eradication have incorporated both sighting records and the relative costs of continued monitoring versus those arising from escape should monitoring be terminated prematurely (Regan *et al.* 2006; Rout, Hauser, & Possingham 2009; Rout, Salomon, & McCarthy 2009). The impact of escape will depend upon how quickly populations increase and spread, which will depend in turn upon how favourable the sites concerned are for the targeted species, as well as the actions and effectiveness of dispersal vectors. This suggests that different stopping rules could apply for populations found in marginal habitats, or in situations where propagules would be poorly dispersed, as compared with situations where the opposite prevails. If eradication is declared, it should be followed up with at least an informal monitoring effort over some years, in order to detect possible resurgences in any populations before these become difficult to manage. For example, the eradication programme targeting *Helenium amarum* in south-eastern Queensland, Australia was declared successful in 2002, but the species was detected five years later in one of its former localities (Panetta & Cacho 2014, table S1).

5.6 Recommendations

Progress towards meeting management goals should be monitored and evaluated both at an individual site level and an overall programme level, and should reflect whether the control operations are performing as desired (the outputs) and what the consequences are to the populations managed (the outcomes). The best indicators vary slightly for containment and eradication, but there are some methods that can be used to assess overall progress (e.g. the eradograph combines progress in terms of extirpating populations and delimiting the invaded area). It is important to determine when to declare success and when to change goals, and there are some general tools to assist in this, in most cases requiring an assessment of the relative costs of different options and the cost of further research. The most important recommendation is to make sure that control efforts are properly monitored (e.g. all control teams should log their locations using GPS devices). Access to monitoring data is essential to the assessment of performance.

Processes	Information requirements	Deliverables
Evaluate management performance at a population level	• Active search effort • Passive detection efforts • Changes in extent over time • Number of individuals controlled (separated into size, age, or reproductive categories) • Source and location of new detections • Response to communication material	• Graph of detections versus search effort over time • An updated risk map • Size or age distribution of numbers controlled year on year • Estimate of rate of seed bank change

(cont.)

Processes	Information requirements	Deliverables
Measure progress towards a management goal at a programme level	• Proportions of populations in different control categories (e.g. active, monitored, extirpated) • Progress scores • Rates of progression and regression	• Graphical representations of progress (e.g. an eradograph)
Decision making	• Programme costs (fixed and variable) • Resources available • List of competing priorities • Defined decision switch points and endpoints	• Recommendation to change resource allocation, if required • Recommendation to continue towards a goal, switch goals, exit coordinated control, or declare goal achieved

6 · *Legislation and Agreements*

Key questions addressed:

- How are alien plants addressed under current legislation and regulation?
- What non-regulatory approaches are available?
- What are the major challenges faced in developing effective legislation, regulation, and agreements?

Legislation plays a critical role in enabling the prevention, detection, and management of plant incursions, i.e. to address the thorns and the thistles. Legislation is required to enable decision making, to frame the risk analysis processes, to determine the roles and responsibilities of those involved, to define pre-border prevention activities, to determine which management measures are acceptable and how they can be applied, and to set conditions for enforcement, compliance, and non-compliance. It can also define the mandates and obligations of the numerous agencies that are involved in incursion response. International and national legislation is required to safeguard a country's biological diversity, protect agricultural and forestry resources, and to ensure access to foreign trade markets. Self-regulation also plays a significant role in managing invasive plants and can take the form of simple voluntary agreements or codes of conduct. However, invasive plants do not respect political or administrative boundaries, creating challenges for coordination. And in the absence of implementation, legislation (formal or otherwise) will not be worth the paper it is written on.

Legislation is defined as the process of making or enacting a law in written form, according to some type of formal procedure, by a branch of government constituted to perform this process (Garner 2009). Legislation and agreements can take many forms, including acts, regulations, treaties, conventions, and policies, and these can apply at international, national, or regional levels. A treaty is an agreement formally signed, ratified, or adhered to between two nations or sovereign countries – an international agreement concluded between two or more states in written form and governed by international law (Garner 2009). A convention is an international agreement between countries comprised of a set of mutually agreed or negotiated terms that have not received the sanction of law and cannot be enforced by the courts. However, conventions often lead to the development of national legislation within a signatory country, as discussed below.

Historically, invasive plant legislation has been reactive, a response to an invasive plant problem that is already established and causing deleterious impacts. However, we are currently witnessing a paradigm shift towards proactive legislation, whereby tools such as spatial predictive modelling and risk analysis are used to predict and prevent the establishment of new weeds.

In this chapter we discuss relevant legislation from international agreements to local by-laws; compare the two major legislative frameworks in which such legislation sits – plant health and biosecurity; explore options for creating lists of regulated species; and then look at some alternatives to legislation. We conclude with a discussion on some of the major challenges to effective legislation.

6.1 Inter-Governmental Treaties, Agreements, and Conventions

It has been estimated that there are about 50 international agreements, and numerous more national and regional strategies, that address some aspect of biological invasions (Shine 2007). These agreements may be binding or non-binding in nature. They provide guidance, standards, agreed rules, and outline general obligations and commitments of signatory partners in responding to biological invasions. Of particular relevance to invasive plant management are the International Plant Protection Convention (IPPC), the Agreement on the Application of Sanitary and Phytosanitary Measures (SPS Agreement), and the Convention on Biological Diversity (CBD) (Lopian 2005).

An example of where more general international collaboration on biological invasions at the policy level has been effective, despite the involvement of multiple nations with different interests, is Antarctica (Box 6.1). While there are some good reasons for this success – the region is isolated with relatively few pathways of introduction (the majority of which are now well understood), and there is a significant appreciation by visitors of the need for compliance – it can nonetheless provide a model for international collaboration.

Box 6.1 *Legislation in Antarctica (Dana M. Bergstrom & Justine D. Shaw)*

Background

The sub-Antarctic islands and the areas of Antarctica that are ice-free can be considered a series of archipelagos, islands isolated either by ice or ocean (Chown, Gremmen, & Gaston 1998; Frenot *et al.* 2005; Terauds *et al.* 2012). While they are small in area relative to the continent as a whole, they support a suite of terrestrial and limnetic communities ranging from polar deserts and permanently frozen lakes to nutrient-rich grasslands and eutrophic ponds. Native plant species range from cyanobacteria, diatoms, bryophytes, and lichens to non-woody flowering plants, with biodiversity increasing from the broad regional categories of continental Antarctica, the Maritime Antarctic (including the Antarctic Peninsula) and the surrounding sub-Antarctic islands. Most biota occurs close to the coast and often coincident with human activities. Biogeographically, the Antarctic continent has at least 15 biologically distinct regions (Terauds *et al.* 2012) and the sub-Antarctic island groups are generally considered separate biogeographic units (Chown, Gremmen, & Gaston 1998).

Native plant taxa range from Gondwanan relics to recent natural arrivals. The mechanism (i.e. by natural means or human-mediated dispersal) by which some newer taxa have arrived is uncertain (Smith & Richardson 2011). Nevertheless, it is clear that over 100 non-native plant species have established in the sub-Antarctic (Frenot *et al.* 2005), with many species becoming invasive, causing substantial damage and fundamental changes to ecosystems. In Antarctica, only two non-native plants *Poa annua* and *Poa pratensis* (now removed; Hughes *et al.* 2015) have established. Both occur in the maritime Antarctic, where the climate is warmer and wetter, and human activity in the form of science and tourism is concentrated (Chown *et al.* 2012). Other species introduced and established in Antarctica have either died out or were removed (e.g. Hughes & Convey 2010).

Pathways of Introduction

Due to the transitory nature of human visitation to and occupation in the Antarctic (hereafter the term 'Antarctic' is used to include both the Antarctic and the sub-Antarctic), pathways of plant introduction have been systematically identified. Cargo, vehicles, equipment, food, personal effects and clothing, ships' hull, and sea chest fouling represent the main vectors for propagules to the Antarctic, with aircraft and shipping providing pathways. Major gateways for propagule transfer to the Antarctic are Hobart (Australia), Christchurch, Dunedin and Bluff (New Zealand), Ushuaia and Punta Arenas (South America), and Cape Town (South Africa). The majority of, but not all, travellers to the region leave from one of these gateways. Ships of national programmes are also known to have cargo loaded in northern ports prior to arriving at a gateway port for more supplies and passengers.

In 2008/2009, the Aliens in Antarctica programme conducted a continent-wide assessment of the threat posed by non-native species, examining clothing and personal effects of around 800 visitors for propagules (Huiskes *et al.* 2014). They found the number of seeds on individuals ranged from 0 to 472, with an average seed load of around ten seeds per person and with scientists posing a greater threat than tourists. However, due to the disparity between tourist numbers (~33 000) and scientists with associated personnel (~7000), overall differences were less. Propagule pressure coupled with amenable climate conditions meant that the Western Antarctic Peninsula was identified as having the highest risk for non-native species establishment (Chown *et al.* 2012). Seeds from 48 families were identified, with the majority of the 2686 seeds found being Poaceae, Asteraceae, and Cyperaceae. Most of the already established non-native plants in the region are from these families (McGeoch *et al.* 2015). The Aliens in Antarctic study provided strong evidence for the need for greater self-regulation (e.g. cleaning personal equipment) and organisation-based regulation (i.e. issuing protocols and conducting inspections to reduce propagule transfer (Huiskes *et al.* 2014)).

Mitigation of the Risk
Most sub-Antarctic islands have management plans that regulate activities and outline measures to reduce non-native plant introductions (de Villiers *et al.* 2005; Shaw 2014). The Antarctic Treaty instils a high degree of connectedness and cooperation between national parties operating in the Antarctic. As such, research that identifies biosecurity threats and measures (e.g. the Aliens in Antarctica programme) has been readily shared through various platforms. For example, the Council of Managers of National Antarctic Programs, in conjunction with the Scientific Committee on Antarctic Research, have produced a checklist for supply-chain managers containing recommendations to reduce the risk of non-native species transfer to the regions (COMNAP/SCAR 2010). This highlights what to look for and how

Figure 1. A biosecurity awareness poster from the Australian Antarctic Programme. Image courtesy of Dana Bergstrom.

Figure 2. A vehicle being hot-washed prior to transportation to Antarctica. Photograph courtesy of Glenn Jacobson.

to search cargo bound for the Antarctic for propagules. The Antarctic Treaty Secretariat hosts an online non-native species manual (ATS 2012) that was developed through their Committee for Environmental Protection. This manual addresses awareness, operational procedures, monitoring, and incursion responses. Many national programmes and the tourist associations have responded to this information in positive ways, such as banning fresh produce and implementing various propagule reduction programmes (Box 6.1 Figs 1 and 2).

Way Forward

With pathways identified and strong mitigation action for reducing propagule pressure quantified, a few areas still require urgent attention. The first is the development of effective, affordable, and novel techniques to reduce propagule pressure in cargo and food; the second is research to develop effective protocols for early detection and rapid responses to incursions; finally, the uptake of high-quality propagule pathway control by all parties involved in Antarctic activities. The Committee for Environmental Protection has recognised the threat from non-native species as one of its highest priorities and as such is actively encouraging uptake of mitigation techniques.

6.1.1 International Plant Protection Convention (IPPC)

The IPPC is an international treaty on plant health that provides the framework for the legislation of invasive pests. The IPPC is the umbrella organisation for phytosanitary activities worldwide on a national, regional, and international level. It provides a framework for international cooperation and has addressed invasive species since its establishment in 1951 (Shine 2007). As of December 2015, there were 182 contracting parties to the IPPC (www.ippc.int/en).

The IPPC is generally considered to be a plant health agreement, but has also been described as an inter-governmental biosecurity network (Gruszczynski 2006). It addresses plant protection both between countries and within a country or region. It is a legally binding international agreement that develops standards for addressing world phytosanitary concerns. The convention is currently described by 23 articles. The first article (1.1) describes its purpose and responsibility: to 'secure common and effective action for preventing the spread and introduction of pests of plants and plant products, and to promote appropriate measures for their control'.

Initially, it was not clear whether the IPPC extended beyond pests of cultivated plants. However, this was clarified at the Fifth Session of the Interim Commission of Phytosanitary Measures in 2003 – the IPPC also considers measures to protect the natural environment and marine flora from invasive plant introductions, as well as invasive plants that might directly or indirectly affect agriculture (Hedley 2004; IPPC 2005; Hewitt, Everett, & Parker 2009).

A key role of the IPPC and its contracting parties is to develop International Standards for Phytosanitary Measures (ISPMs) for safeguarding plant resources. Adopted ISPMs address issues such as surveillance, pest management, **quarantine**, treatment, export certification, and risk analysis. There are four types of standards: reference, pest risk analysis, concept, and specific. The IPPC has developed ISPMs that are significant in any discussions pertaining to the legislation of invasive plants, for example ISPM No. 2 *Framework for Pest Risk Analysis* (IPPC-FAO 2007) and ISPM No. 11 *Pest Risk Analysis for Quarantine Pests* (IPPC-FAO 2013b). The ISPMs strive to provide a consistent, science-based process that can be used to arrive at a regulatory decision. ISPMs are also designed to harmonise phytosanitary measures applied in international trade and to protect biodiversity by preventing the movement of invasive plant pests (FAO 2012).

Article IX of the IPPC makes provisions for the formation of Regional Plant Protection Organizations (RPPOs). There are currently nine RPPOs and each has its own programme and activities. To date, the most active RPPO with respect to invasive plants has been the European and Mediterranean Plant Protection Organization (EPPO) (Box 8.2), while the North American Plant Protection Organization has developed a regional standard for 'Pest Risk Assessment for Plants for Planting as Quarantine Pests', published in 2008 (RSPM No. 32). The IPPC has also directly led to the creation of plant health units at the national level via the creation of National and Regional Plant Protection Organizations (Section 6.3).

6.1.2 Agreement on the Application of Sanitary and Phytosanitary Measures (SPS Agreement)

The SPS Agreement is in place to ensure that global commercial trade is not hindered by artificial barriers or protectionism, and that **phytosanitary measures** are based on international standards. It has been in force for all countries since 2000 (Devorshak 2012), and is an agreement under the World Trade Organization (WTO). Articles 2 and 5 of the SPS Agreement are particularly relevant to developing any legislation for invasive plants. Article 2 states that all Members have the right to adopt phytosanitary measures to protect against the introduction of invasive pests and that these measures must be based on science. A pest risk assessment as per the IPPC's ISPMs, for example, is one way measures can be scientifically justified. It is the importing country's obligation to demonstrate risk and to justify any SPS measure applied. Article 5 requires that phytosanitary measures are based on a risk assessment underpinned by the most appropriate scientific evidence.

The SPS Agreement recognises the IPPC as the only acceptable international standard-setting body with regard to plant health.

6.1.3 United Nations Convention on Biological Diversity

The United Nations Convention on Biodiversity (CBD) is an international legally binding treaty for the protection of biological diversity, and entered into force in 1993. As of December 2015, 196 parties have ratified the treaty, including all members of the United Nations except for the USA (www.cbd.int). The CBD makes provisions for addressing invasive species that are also included in many programmes established by the CBD's Conference of the Parties (COP). In particular, Article 8(h) states that each contracting party shall 'prevent the introduction of, control or eradicate those alien species which threaten ecosystems, habitats or species'. It is the only global instrument that mandates prevention and mitigation measures for all invasive species (Shine, Williams, & Burhenne-Guilmin 2005). The COP further recognised that there is an urgent need to address the impact of invasive species through a three-stage hierarchical approach, where priority is given to preventing new introductions, early detection, and rapid response, followed by eradication, containment, and long-term management efforts (Shine, Williams, & Burhenne-Guilmin 2005). At the COP meeting in April 2002, 15 non-binding guiding principles were adopted to assist governments and organisations in developing effective strategies to minimise the spread and impact of invasive species. The guiding principles are categorised into three groups: (1) general, (2) prevention, and (3) introduction of species. In 2010 in Aichi (Japan) the CBD Parties approved the following:

By 2020, invasive alien species and pathways will be identified and prioritised, priority species will be controlled or eradicated, and measures will be in place to manage pathways to prevent their introduction and establishment. (Aichi Target 9)

While the CBD is concerned with organisms that impact biological diversity, and the IPPC is concerned with pests that impact plant health, there is some overlap in scope when it comes to invasive species (MacLeod *et al.* 2010). The IPPC has gone as far as adding an appendix to ISPM No. 5 which discusses similarities between the two agreements, and the two groups work together through working groups. In all cases legislation is needed at the national level for the principles of the CBD and IPPC to come into effect.

While the CBD has arguably led to an increase in the number of countries with national policies relevant to biological invasions (McGeoch *et al.* 2010), the focus of the IPPC and the commercial implications of the SPS Agreement have likely had a bigger impact on national legislation and on the prevention of biological invasions.

6.2 Legislation at the National Level and Below, Policies, Acts, and Regulations

Governments create legislation to address all stages of the invasion continuum (see Fig. 1.2): to prevent the entry of potential invaders from other countries pre-border; to rapidly respond and eradicate new incursions; to slow the spread of invasive plants within domestic borders; and to mitigate the impact and protect assets if spatial distributions become too large.

Governments have been battling invasive plants through legislation for decades, or as Evans (2002) so aptly described it, declaring a 'war on weeds' (see Timmons (1970) for a review of early weed control and legislative efforts in Canada and the USA, and Appleby (2005) for an update). Historically, governments have created legislation to protect agricultural interests, and more recently to protect trade, ecosystem services, and biodiversity. However, governments also want to stimulate economic development and foster international trading relationships. Therefore, the role of government can become complex and conflicting, for example working to keep an invasive plant outside its borders, while maintaining and developing trading relationships with countries whose commodities might be contaminated with the invasive plant of concern. There are, unfortunately, plenty of examples where one branch of government is busy trying to control an invasive species, while in a neighbouring office other officials are promoting the import and use of the same species for agriculture (Paynter *et al.* 2003; Driscoll *et al.* 2014). It is evident that the public policy environment and the role of government in creating effective legislation are complex and challenging with regard to invasive plants. An additional layer of complexity is that there is also often legislation and regulation at a level below national governments. This can, of course, mean that regulations are tailored more specifically to the risks posed, but it can also create conflicts in policy.

A policy at a national level is the general principle by which a government is guided in its management of public affairs (Garner 2009). Recently, we have seen many countries publish policy statements on invasive species and some more specifically on invasive plants. For example, in 2012 the Canadian Food Inspection Agency (CFIA) published an Invasive Plants Policy that states that it will regulate invasive plants in the same fashion as any other pest and that decisions to regulate will be based on risk analysis and pathways of introduction. In the USA, the presidential Executive Order 13112, published in 1999, established the government's policy to prevent and control invasive species by establishing a federal Invasive Species Council, as well as an Invasive Species Management Plan. In 1998, Biosecurity New Zealand published a Policy Statement on Unwanted Organisms, which clarifies the responsibilities of various officers involved with unwanted organisms in New Zealand. Policies can also be the way in which regulations will be enforced at a local level.

6.2 Legislation at the National Level and Below · 149

CAP.XL.

An Act to prevent the spreading of Canada Thistles in Upper Canada

(Assented to 18th *September,* 1865.)

HER Majesty, by and with the advice and consent of the Legislative Council and Assembly of Canada, enacts as follows:

1. It shall be the duty of every occupant of the land in Upper Canada, to cut, or to cause to be cut down all the Canada thistles growing thereon, so often in each and every year as shall be sufficient to prevent them going to seed; and if any owner, possessor, or occupier of the land shall knowingly suffer any Canada thistles to grow thereon and the seed to ripen so as to cause or endanger the spread thereof, he shall upon conviction be liable to a find of not less than two or more than ten dollars for every such offence.
2. It shall be the duty of the Overseers of Highways in any Municipality to see that the provisions of this Act are carried out within their respective highway divisions, by cutting or causing to be cut all the Canada thistles growing on the highways or road allowances…
3. It shall be the duty of the Clerk of any Municipality in which Railway property is situation, to give notice in writing to the Station Master of said Railway resident in or nearest to the said Municipality require him to cause all the Canada thistles growing upon the report to the said Railway Company with the limits of the said Municipality to be cut down ….
4. Each Overseers of Highways shall keep an accurate account of the expense incurred by him in carrying out the provisions of the preceding sections of the Act, ……
5. The Municipal Council of the Corporation shall cause all such sums as have been so paid under the provisions of this Act, to be severally levied on the lands described in the statement of the Overseers of Highways…
6. Any person who shall knowingly vent any grass or other seed among which there is an seed of Canada thistle, shall for every such offence, upon conviction, be liable to a find of not less than two nor more than ten dollars.
7. Every Overseer of Highways or other officer who shall refuse or neglect to discharge the duties imposed on him by this Act, shall be liable to a find of not less than ten more than twenty dollars.
8. Every offence against the provisions of the Act shall be punished, and the penalty hereby enforced for each offence shall be recovered and levied, upon conviction, before any Justice of the Peace; and all fines imposed shall be paid into the Treasury of the Municipality in which such conviction takes place.

Figure 6.1. There has been a long history of acts dedicated specifically to invasive plant control, e.g. the 1865 *Canada Thistle Act of Upper Canada.*

An act in the law creates, transfers, or extinguishes a right and usually provides the basis for the exercise of legal power (Garner 2009) (e.g. Fig. 6.1). As such, an act generally sets out the framework of a regulatory scheme and delegates the authority to develop details and regulations, and consequently is often referred to as primary legislation. Relevant examples are the *Plant Protection Act*, the *Seeds Act* (Canada); the *Plant Protection Act* and the *Federal Seed Act* (USA); and the *Biosecurity Act* (New Zealand). In the USA, the legal authority for the US National Weed Program and the strategic rationale for programme management and goals is the *Plant Protection Act*. Signed into law in 2000, it replaced the *Federal Noxious Weed Act* of 1974. The *Environmental Protection and Biodiversity Conservation Act* of 1999 and *the Quarantine Act* of 1908 regulate the importation of plants into Australia. Most acts provide provisions to create lists or schedules of invasive plants, weeds and their seeds by way of regulations.

Regulations enable additional detail to be established in law for particular sections of acts where it is not appropriate to include such detail in the act, or to establish species-based programmes or to prepare lists of species to which the act applies. Regulations are a form of law, often referred to as delegated or subordinate legislation. Like acts, they have binding legal effect and usually are to be applied generally, although there can be explicit provisions to deal with specific situations. For example, under the new *Queensland Biosecurity Act 2014*, a regulation may establish a biosecurity zone for a particular pest or a group of pests of a particular crop. The authority to make regulations can be expressly delegated by an act (i.e. an enabling act) to a minister or administrative agency, although the exact process of approval of regulations will vary across different legislative systems. Regulations generally take less time to establish than acts (e.g. 1–2 years vs 3–4 years), but can take a very long time if there are contentious issues or if bureaucratic systems are inefficient. Regulations may identify the invasive plants that a country believes constitute a risk to its plant health, how they might enter a country, and where measures should be implemented (Hedley 2004). For example, the United States Department of Agriculture (USDA) has published *Noxious Weed Regulations* under its *Plant Protection Act*, while South Africa published *Alien and Invasive Species Regulations* in 2014 under its *National Environmental Management: Biodiversity Act*.

While it is encouraging to see there has been a steady increase in the number of countries with a national policy relevant to biological invasions (McGeoch *et al.* 2010), it is much less clear whether such legislation has been effective. Poorly written acts, regulations, and policies can hamper control efforts, not least by creating a distrust of government processes and actions.

6.3 Plant Health and Biosecurity

There are two general frameworks under which invasive plants are addressed – plant health and biosecurity. Both approaches strive to prevent invasive plant introductions, and when prevention efforts fail, to guide the application of management efforts or phytosanitary measures. Both frameworks lead to legislation. Each approach places emphasis on different aspects of invasive plant control, but they have the same end goal.

In a policy context, plant health is considered the first line of defence against invasive plants, as it can implement phytosanitary measures on imports and the entry of plants and plant products at border crossings. Traditionally, plant health has focused on pests of plants, or those pests that have impacted managed agricultural, forestry, and horticultural systems. The responsibility therefore often lies with government departments concerned with agriculture. However, during the 1990s the scope of plant health expanded to address plants as pests, which can be invasive plants, and their impacts on the natural environment (MacLeod *et al.* 2010).

Article IV of the IPPC requires contracting countries to create an official National Plant Protection Organisation (NPPO) to discharge the functions of the IPPC. The NPPO is responsible for issuance of phytosanitary certificates, delivering detection surveys and inspections, and conducting pest risk analyses. A country's NPPO implements the IPPC following a hierarchical approach comprising: (1) prevention of the introduction of the pest as the preferred measure; (2) eradication at the earliest possible stage of invasion; and then (3) containment and control if the previous measures fail or are not feasible.

One future opportunity is for the IPPC and NPPOs to extend their efforts and consider instigating plant protection organisations within a country. Within a country, potential synergies exist with current state or provincial government programmes, both administratively and legislatively. This would create a public policy environment for efforts to harmonise legislation. For example, many state or provincial governments that list noxious weeds do not employ a science-based approach, but the frameworks for risk analysis developed by the IPPC could be adopted.

The tools employed in plant health systems to address invasive plants include the use of risk analysis, import permits, phytosanitary certificates, pre- and post-border inspections, regulated pest lists, surveillance systems, reporting, and enforcement. For example, the requirement for an import permit is important as it allows a country to review applications and screen for possible invasive plants. In most cases, if something new is being imported into a country the importer requires an import permit and the application process triggers the risk analysis process. All of these tools have a focus on preventing introductions of invasive plants and have a decreasing role as the spatial distribution of the plant increases within a country.

In contrast to plant health, the term biosecurity has traditionally been associated with infectious diseases, biological weapons (Armstrong & Ball

2005), or with bioterrorism and homeland security (Burnette *et al.* 2013). However, it has been extended to pest control such that biosecurity can be regarded as the rebranding of a centuries-old practice of battling agricultural pests (Waage & Mumford 2007). The FAO notes that biosecurity includes: (1) frameworks that set out the broad course of action to address risk; (2) legislation (i.e. laws and regulation) defining appropriate powers to act or that set out an enabling environment; (3) a requirement for an institutional framework, clear policy and legal frameworks, and scientific capacity to implement risk analysis; and (4) many different kinds of stakeholders, generally branches of government at a national level with responsibilities for food safety, public health, agriculture, forestry, trade, fisheries, or the environment (FAO 2007).

The term biosecurity has been used to address various aspects of invasive plant management, including discussions on risk assessment tools (Pheloung, Williams, & Halloy 1999), identification tools (Armstrong & Ball 2005), the international trade of plants (Brasier 2008), approaches to addressing non-native organisms (Meyerson & Reaser 2002), agriculture (Waage & Mumford 2007), and economic benefits (Kriticos, Phillips, & Suckling 2005). Recently, we have seen entire books dedicated to biosecurity and invasive species (e.g. Dobson *et al.* 2013). In New Zealand and Australia, the term biosecurity is associated with concerns for native flora and fauna within an environmental conservation ethic impacting trade and export, and in Britain and Europe biosecurity focuses on agricultural pests and diseases (Barker, Taylor, & Dobson 2013).

New Zealand probably has the most comprehensive biosecurity legislation in its *Biosecurity Act* of 1993 (Meyerson & Reaser 2002; Armstrong & Ball 2005). The *Biosecurity Act* aims to protect primary industries and the environment by unifying all previous pest management into one single piece of legislation with a central authority (the Ministry for Primary Industries). The *Biosecurity Act* provides the legal basis for excluding, eradicating, and managing pests and unwanted organisms in New Zealand, including invasive plants (Yamoah, Gill, & Massey 2013). It makes provisions for regional councils that provide leadership in activities to prevent, reduce, or eliminate the impacts of invasive plants. The *Biosecurity Act* defines an unwanted organism as any organism that a chief technical officer believes is capable or potentially capable of causing unwanted harm to any natural or physical resources or human health. Under section 52 of the *Biosecurity Act*, it is prohibited to release or spread any pest or unwanted organism in New Zealand. The *Hazardous Substances & New Organisms Act* 1996 also provides legislative authority to address invasive plants. The *Biosecurity Act* addresses invasive plants that might be introduced with imported goods, while the *Hazardous Substances & New Organisms Act* deals with those where a permit to import is required.

In Australia, a new *Biosecurity Act* came into effect on 16 June 2016, replacing the *Quarantine Act* of 1908. It will be jointly administered by the Australian Agriculture and Health Ministers. The Australian Government Department of Agriculture and Water Resources undertakes a range of science-based risk assessments (including import risk analyses) and provides biosecurity policy advice to protect Australia's human, animal, and plant health status and natural environment. Plant Biosecurity, within the Australian Government Department of Agriculture and Water Resources, regulates the importation of new plant species, following a decision-making policy consistent with relevant international treaties. All proposed new plant imports are assessed for invasive potential, in addition to examining the risk that they will carry exotic diseases or insects, before they are permitted entry. Plant Biosecurity uses a pre-entry weed risk assessment system for assessing all new plant imports (Pheloung, Williams, & Halloy 1999), which was implemented in 1997 (Weber *et al.* 2009). This system is now used on all new plant imports, whether they enter Australia as seeds, nursery stock, or tissue culture and regardless of their end use in Australia. The Department of Environment was involved in developing the weed risk assessment system and accepts the outcomes in its legislation (the *Environmental Protection and Biodiversity Conservation Act 1999*). Plant Biosecurity maintains a record of all weed risk assessments.

The responsibility for biosecurity (as for plant health) often lies with government departments concerned with agriculture, with a few notable exceptions (e.g. in South Africa it is housed with the Department of Environmental Affairs). However, regardless of the institutional home of the relevant act and any resulting regulations, close interactions between a range of government departments (also including health and trade) are needed: 'Building a biosecurity system is a collaborative project. It takes a whole country' (Biosecurity New Zealand 2011).

6.4 Lists

A common approach in many acts and regulations is to define different categories of invasive plants (or weeds) of concern and produce lists. These can be for regulatory or non-regulatory purposes. Regulatory lists identify the regulated invasive plant of concern and are linked to the relevant act, regulation, or policy which provides the legislative authority.

Under the IPPC, for example, NPPOs are required to maintain lists of regulated pests in order to communicate quarantine measures to other countries. Hence lists become an important inter-regional communication tool (e.g. IPPC-FAO 2003).

From both a biosecurity and a plant health perspective there are three main types of lists – permitted lists (also referred to as white lists); prohibited lists (black lists); and lists of plants requiring risk analysis (grey lists). Lists can be declared through a Ministerial Order, separate from legislation, to allow faster response to new species (but this process is susceptible to political decisions). The ability to develop and enforce such lists will depend on the context. Permitted lists are feasible only if an area has a small number of entry points that can be controlled and where the risks are well understood. Prohibited lists are much easier to manage, but conflicts can quickly arise if taxa are included that have value to some stakeholders. The use of lists requires that species can be identified. This can be problematic, particularly if taxonomic boundaries are blurred due to hybridisation, or if identification requires specialised taxonomic skills that are not readily available.

As an example, Australia adopted a permitted list in 1998 (Schedule 5 of the Quarantine Proclamation 1998) which replaced a prohibited list. Permitted species can be imported and sold, while all other non-listed species are automatically prohibited until a weed risk assessment is completed. To be permitted, a species must either be already present in Australia and not under official control or have been assessed using the weed risk assessment system as having low potential to become a weed in Australia. Permits for plants are only issued at the species level. This system is used on all new plant imports, whether they enter Australia as seeds, nursery stock, or tissue culture and regardless of their intended end use. The implementation of the weed risk assessment system is a component of the Australian Weed Strategy. The use of a permitted list and a weed risk assessment to determine what can be imported represented a significant policy shift for regulating weeds.

Regulatory lists are also employed at regional levels (i.e. state or provincial), often called noxious weeds lists. Recently we have seen these lists being further categorised based on spatial distributions and expanded to include non-agricultural weeds. We have also seen a shift in the regulatory focus towards invasive plants that are not yet in the area. For example, in Canada the provinces of Alberta and Saskatchewan have both revised their weed legislation and placed an emphasis both on addressing invasive plants not yet present in the provinces and on those early in the invasion process. In 2010 a revised Alberta *Weed Control Act* categorised weeds into two groups: (1) prohibited noxious – are not yet found in Alberta or are found in isolated populations that have potential to be eradicated before they become widely established; and (2) noxious – eradication from the region is no longer feasible, but control efforts are needed to prevent further spread. The Saskatchewan *Weed Control Act* of 2010 created the following categories: (1) prohibited – plants that are absent or rare, but pose a significant threat; (2) noxious – present in isolated areas, containment not possible, and spread likely; and (3) nuisance – weeds that are widespread. In the past the regulatory focus, and hence the management focus, has been on agricultural weeds already present in an area and causing economic damage; authorities are beginning to categorise lists, shifting the focus towards prevention.

The importance of using a science-based risk analysis approach to creating lists is obvious, but this is not always adopted. In many areas regional governments decide which plants are invasive by weight of evidence. In some cases it is by committee or by councils who have seen the plant spreading and have received calls from concerned citizens. The deleterious risks associated with introductions of invasive plants or commodities that might be contaminated can be elucidated through risk assessment (Section 2.4). The IPPC has published standards on how to conduct a risk analysis and what elements should be evaluated, namely: ISPM No. 2: *Guidelines for Pest Risk Analysis*; and ISPM No. 11: *Pest Risk Analysis for Quarantine Pests*.

Additional lists are created in order to target surveillance activities. Alert lists, watch lists, or early detection and rapid response (EDRR) lists will typically consider some combination of the likelihood of a species being introduced, whether the environment is favourable for that species, and the probability of invasion or impact if it is introduced to a suitable area (Faulkner *et al.* 2014). If a species that is on such a list is detected, this should be an immediate trigger for all of the other activities associated with managing an incursion. One approach to producing a watch list is shown in Fig. 6.2.

6.5 Alternatives to Legislation

Voluntary agreements between stakeholders are an attractive functional alternative to formal legislation. These instruments involve a voluntary and moral commitment from an organisation to implement an agreement (Halford *et al.* 2014). For governments, voluntary agreements can save

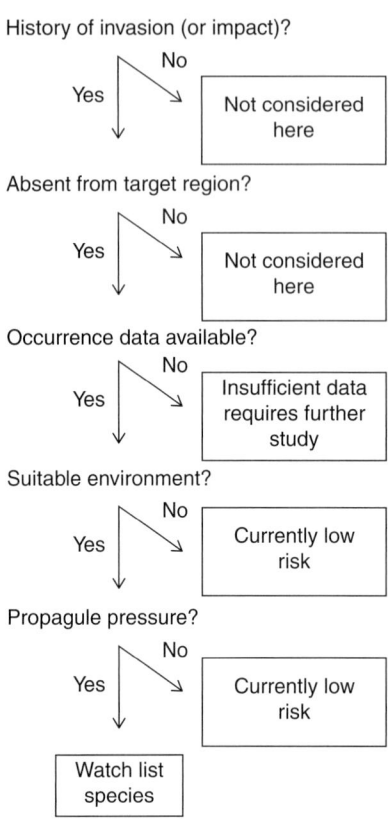

Figure 6.2. A simple decision tree for creating a watch list. One of the most reliable predictors of invasiveness for a species in a new region is whether it has been documented as invasive elsewhere. Of course this feature will be of no assistance if the species has no history of introduction beyond its native range. Redrawn from Faulkner *et al.* (2014) with permission.

on compliance, administration, and other management costs (Segerson & Miceli 1998), and for industries and other stakeholders it can mean they have a much more direct say on the issue of invasive plants, resulting in rules that are more appropriate to the risk posed. In general, self-regulation involves developing sets of broad guidelines (e.g. codes of conduct) with the objective of increasing consumer confidence or resolving conflict.

One of the earliest examples is the St. Louis Declaration on Invasive Plant Species. Following a three-day workshop in 2001, voluntary codes of conduct were drafted with a set of guidelines for each of several target groups: government, nursery professionals, gardeners, landscape architects, botanic gardens, and arboreta. The Belgian Code of Conduct, launched in 2011, is an example of a carefully managed consultation process that included a series of ten meetings between horticultural professionals and scientists (Halford *et al.* 2014). The code itself consists of five 'good' practices: (1) keep informed about the list of invasive plants in Belgium; (2) stop stocking specific invasive taxa; (3) disseminate information on invasive plants to consumers; (4) promote the use of non-invasive alternatives; and (5) take part in early detection activities. The code of conduct has two key lists: a consensus and a communication list. Horticultural professionals developed a consensus list of invasive plants that can be withdrawn from sale. The communication list comprises invasive plants that can be traded, but when consumers purchase such plants they are advised that planting in certain areas has the potential for the invasion of sensitive habitats. Developing the lists involved many difficult discussions between the horticultural professionals and scientists (especially on which plants were invasive), and, while the authors felt the approach had value, the small number of invasive plants that had been withdrawn from sale (i.e. the consensus list) suggested that there needed to be a better engagement between the horticultural industry and invasive plant managers. Another notable example is the Council of Europe and the European and Mediterranean Plant Protection Organization's (EPPO) Code of Conduct on Horticulture and Invasive Alien Plants.

These codes of conduct are excellent examples of proactive partnerships where the horticultural industry has taken the lead, effectively becoming the champion in implementing change and raising awareness. There is certainly motivation and willingness within the horticultural industry to self-regulate, as opposed to being subject to formal regulation. However, there are some fundamental problems with self-regulation in that it leaves it up to a specific industry to identify the relevant risks (Reichard *et al.* 2005). Moreover, despite years of attempts at self-regulation many nurseries are either not willing to comply or are not aware of the codes of conduct at all (Burt *et al.* 2007). This is most likely because of a lack of a communication strategy and the overall size and spatial distribution of the industry, as opposed to the industry turning a blind eye. As Drew *et al.* (2010) pointed out, the horticultural industry is highly decentralised and characterised by a complex market chain. However, the development of codes of conduct demonstrates an acknowledgement of the problem and the industry's willingness to be part of the solution.

Accords are similar to codes of conduct in that they are simple agreements to work together. The New Zealand Plant Pest Accord is an example of a cooperative agreement between the Nursery and Garden Industry Association, regional councils, and government departments with biosecurity responsibilities to work together in the common interest of addressing invasive plants. It was created in 2001 to help prevent the spread of invasive plants through the nursery trade pathway and in particular to ensure proper consultation and discussion. As of July 2012, the members of the New Zealand Plant Pest Accord have agreed to add 13 plants as pests that are illegal to propagate, distribute, or sell under the *Biosecurity Act 1993*.

It is also important to develop and sustain a culture of biosecurity. Examples include farmers not accepting forage shipments without guarantees that these products are free of harmful contaminants, landowners not allowing vehicle entry without prior cleaning, boat-owners being provided with areas to wash down boats before transport to other water bodies, and community education initiatives to encourage gardeners to dispose of garden waste responsibly. Where such measures are linked to benefits, e.g. clean-down is required for access to an area, they can be very effective in reducing spread.

6.6 Challenges to Legislation and Agreements

There are many challenges to legislation and agreements regarding invasive plants, and a few will be discussed below.

6.6.1 Reactive Rather Than Proactive

Legislation has generally been developed reactively, in response to public or economic concerns over an invasive plant, usually well after the plant has established and incurred deleterious environmental and economic impacts, i.e. we are still in catch-up mode. More recently, however, we have seen a paradigm shift where authorities are using risk analysis tools, spatial predictive modelling, and environmental scanning exercises to identify potentially invasive plants that are not yet present in a region, and then developing proactive legislation to prevent those introductions or imports.

6.6.2 Reconciling Conflicting Policies

World economies strive for increased trade liberalisation, but these factors challenge national regulatory authorities and border protection systems

(Burgiel et al. 2006). International agreements such as the SPS Agreement, the North American Free Trade Agreement, and the General Agreement on Tariffs and Trade seek to liberalise trade, which can lead to more unintentional introductions of invasive species; the macroeconomic effects have yet to be examined (Jenkins 1996). Campbell (2001) notes that the SPS Agreement was written in 1994 by authorities who were interested in promoting trade and were not well versed in invasion ecology.

6.6.3 Fragmented Legislation

Invasive plants do not respect international, national, or domestic political borders, rather following biogeographic boundaries or patterns associated with specific human activities (Rouget et al. 2015). Hence, preventing the spread of invasive plants requires cooperation and communication across multiple jurisdictions and authorities. The challenge is that each jurisdiction or authority has its own set of mandates, legislative responsibilities, and more importantly, available resources. As such, legislation addressing invasive plants within a country is often fragmented across many different acts and regulations. Hence, there are frequent calls for the harmonisation of legislation, both nationally within different levels of government and internationally between countries. For example, in Europe there are as many as 29 domestic Acts, Orders and Regulations with relevance to invasive species (DEFRA 2003), while Corn and Johnson (2013) noted that there is no comprehensive legislation or single law that provides coordination among authorities in the USA with respect to invasive species. There is often no coordinating framework between the various authorities to ensure cooperation and coordination across the different political sectors (Wittenberg & Cock 2001; Meyerson & Reaser 2002). In most cases, as stated above, this is a result of the different political mandates and obligations of the legislative authorities. As far as we know, the influence of this bureaucratic decision on the effectiveness of the management of incursions has not been studied, but would likely be a fruitful area of policy research. Invasive plants are generally the responsibility of numerous political or administrative authorities, each with its own specific mandates, and as a result harmonisation might not be achievable.

6.6.4 e-Commerce

The challenges to regulatory agencies from e-commerce are multifaceted: (1) most invasive plants are available for sale somewhere; (2) national legislation does not apply to foreign exporters; (3) e-commerce vendors often provide no physical address; (4) when shipped, plants are often mislabelled; and (5) in many cases traders have little or no appreciation of the problem. For example, the pop-singer Katy Perry produced an album (*Prism*) with limited editions that contained seed-paper. While a nice gimmick, it had the potential to spread invasive plants and so the limited edition was banned by the Australian Quarantine and Inspection Service. Of more direct concern was a study by the Minnesota Department of Natural Resources that found internet and catalogue businesses in the USA offered invasive plants for sale that were on state and federal restricted invasive plant lists (Perleberg 1999). Maki and Galatowitsch (2004) found that State and Federal prohibited species in the USA could be acquired 92% of the time they were ordered. Thum, Mercer, and Wcisel (2012) found that some invasive plants are often sold on the internet with incorrect names, having both ecological and economic consequences. Kay and Hoyle (2001) examined ornamental catalogues and websites, finding that most species of aquatic and wetland invasive plants listed either as federal noxious weeds or as noxious weeds in one or more US states were available on internet sites that were selling invasive plants worldwide. They suggested that a major challenge and serious complicating factor is that many regulatory agencies, at both the federal and provincial level, are unfamiliar with noxious aquatic and wetland plants, particularly if they are small and are mixed with other larger plants.

Practical recommendations to reduce the risks associated with e-commerce include developing a targeted education and awareness programme aimed at e-commerce vendors, initiatives to identify and monitor high-risk vendors, mandatory labelling requirements, and new international standards for e-commerce.

6.6.5 Dealing with Conflict Species

As discussed previously, there is often one group that benefits from introduced plants, while the costs of an invasion might be borne at a later stage by different stakeholders far away from the initial sites of introduction. Introduced plant species are used for a plethora of different reasons, e.g. in the aquarium plant trade, for pasturelands, for horticulture use and landscaping, for soil conservation and stabilisation, for biofuels, and for forestry. In each case there is the potential that the introduced species will subsequently become invasive. Unless those benefiting from the plants accept that the benefits they accrue are not worth the risks involved, then coordinated control will be close to impossible.

This issue is not helped by the fact that in many cases countries have conflicting legislation – regulations that prevent the introduction and spread of invasive plants, as well as legislation that promotes their use. This is not intentional, but can occur when two different public policy streams mature and react independently to their respective socio-economic environments. As Auld and Johnson (2014, p. 4) note 'the problem of widely planted pasture species invading natural ecosystems poses a complex problem (or "ambiguity") requiring collaborative policy development amongst organisations and, likely, a range of prioritisation, risk assessment and weed management tools'. An important recent example of the challenges posed to regulators by species where there are conflicts of interest is the use of invasive plants as a biofuel feed-stock for renewable energy (Box 6.2).

Box 6.2 *Regulating the Use of Potential Invaders for Bioenergy (Lauren D. Quinn)*

Globally, there is a push to develop bio-based fuels that will reduce greenhouse gas emissions relative to fossil fuels. Biofuel products (e.g. ethanol, diesel, combustible solid biomass, syngas, and others) are made from a wide variety of materials, from traditional food crops (e.g. corn and soy), to municipal or agricultural green waste, to animal manure. However, a recent emphasis has been the development of 'second generation' cellulosic feedstocks that have the potential to be converted to ethanol (Sorda, Banse, & Kemfert 2010). As they are not derived from food crops, which often require intensive irrigation and fertiliser application, second-generation feedstocks can be grown on less valuable 'marginal' land and preserve national food supplies. In addition, crops developed as second-generation feedstocks grow rapidly and produce extremely large amounts of biomass, and many are perennial, display C_4 photosynthesis, and possess other desirable traits. However, it has been pointed out that many of these traits are strongly correlated with invasiveness (Raghu *et al.* 2006). Moreover, several known invaders have been proposed or are currently being grown as cellulosic feedstocks in North America and around the world (Quinn 2014; Quinn *et al.* 2014b).

In the USA, the 2007 *Energy Independence and Security Act* established the current Renewable Fuel Standard (RFS), which mandates that an increasing portion of the transportation fuels sold in the USA must be derived from non-food-based renewable sources, including second-generation cellulosic feedstocks. Under the RFS, fuel producers petition the US Environmental Protection Agency (EPA) to approve new fuel sources ('pathways') for eventual sale. According to the regulations as written, the EPA evaluates these new fuels according to their greenhouse gas emissions profiles and is not required to consider invasion potential. The majority of the ethanol sources that have been approved by the EPA thus far have little to no invasion potential (e.g. crop residue, forestry thinnings, municipal solid waste), but in January 2012 the EPA approved *Arundo donax*, *Camelina sativa*, 'energy cane' (*Saccharum* spp.), and *Pennisetum purpureum*. All of these species are known invaders in the USA and other parts of the world. For example, *A. donax* is a widespread invader in riparian areas in California and Texas, where it causes significant negative environmental and economic impacts to resident species, fire regimes, and water supplies, and it appears in numerous 'worst of' lists, including a list of 100 of the world's worst invaders (Lowe *et al.* 2000). The EPA's decision to approve these feedstocks caused such a loud and immediate uproar from the environmental community that the EPA withdrew its ruling two months later pending further public comment and evaluation. Eventually, the EPA approved all four species based on their greenhouse gas emission profiles, but stipulated that anyone producing fuel from *A. donax* and *P. purpureum* must show non-significant risk of spread or supply a risk mitigation plan that requires recordkeeping, use of non-specific best management practices, and third-party inspections.

Recently, the EPA has consulted the US Department of Agriculture (USDA) and other experts to weigh invasion concerns in its determinations for new feedstocks, but unless invasiveness changes a feedstock's greenhouse gas profile, invasiveness alone is unlikely to alter the outcomes of these assessments. Another problem is that a number of feedstocks have been petitioned or approved under a common name (e.g. 'energycane') or a genus name (*Miscanthus*), meaning that any future varieties could potentially be subsumed under these names, even if they possess traits that confer greater risk of invasion.

Currently, the USDA is the only other federal agency that regulates bioenergy plantations. Its Biomass Crop Assistance Program allows for subsidies for plantations of perennial grasses, including *Miscanthus* × *giganteus*. Unlike the EPA's new fuel pathway approval process, the Biomass Crop Assistance Program specifically exempts invasive or potentially invasive feedstocks from receiving subsidies and provides for more specific management practices designed to prevent spread even by low-risk non-native feedstocks.

A few US states are regulating bioenergy plantations with respect to invasion concerns. Florida was first, with a permit and bond requirement for non-native bioenergy plantations larger than two acres, and an exemption for noxious weeds and invasive plants, defined as 'naturalized plants that disrupt naturally occurring native plant communities'. Mississippi followed with a very similar statute in 2012, but it did not define the terms 'invasive', 'non-native', or others. Oregon has established a state-wide control area for the production of *A. donax* for biomass, requiring permits, bonds, and specific management practices. Maryland's Renewable Portfolio Standard disqualifies invasive exotic plant species (not defined) to be used for biomass.

There are several problems with these regulations. First, only four states currently recognise the potential for non-native bioenergy plantations to escape cultivation and become invasive, but the bioenergy industry operates on a much larger geographic scale. Second, most of the regulations fail to adequately define key terms relating to invasiveness, which could lead to confusion, misrepresentation, and delay in permit approval (Quinn *et al.* 2014b). Some states refer to their noxious weeds laws in defining invasion terms, but a recent study showed that, on average, state noxious weeds laws only list about 20% of the invaders that impact natural areas (Quinn *et al.* 2013). Finally, those states that require bond payments set unrealistic prices for the removal of wayward feedstocks. In Oregon, the bond payment for *A. donax* plantations is USD 100 per acre, but it has been shown that it can cost up to USD 25 000 per acre to control *A. donax* and restore resident vegetation in California (Giessow *et al.* 2011). In general, the language in these regulations requires bonds to cover only removal of the crop from the cultivated area, not beyond.

Clearly, current state and federal regulations are inadequate to prevent invasions by novel bioenergy crops. My colleagues and I have recommended a number of regulatory reforms that would address these issues. While the current cadre of EPA employees is proactively and admirably seeking out invasiveness data in its handling of new fuel pathway petitions, it will be important to include this as a formal requirement in revised EPA regulations. And it will be important to provide a means of assessing invasiveness. Weed Risk Assessments can be used to determine invasion potential in bioenergy crops (Barney, Smith, & Tekiela 2014), and should be specified as part of the assessment process in revised state and federal policies (McCubbins et al. 2013; Quinn et al. 2013; Quinn et al. 2014a; Quinn et al. 2014b). In addition, agencies could rely on 'white lists', or pre-approved lists of low-risk crops, in approving or denying novel feedstocks for production. For states that require bond payments, these should be revised to account for realistic costs associated with removal of plants within *and beyond* cultivated fields. As an alternative to bonds, biofuel producers could be held liable for negligence if invasions occur as a result of high-risk crop introduction. Most of the existing state and federal regulations do not specify in enough detail the management practices that would prevent escape. Specific plans should be included for prevention, monitoring, spot-removal, and notification of authorities when larger escaped populations are detected (Barney 2012; Quinn et al. 2014b).

There are various options to deal with conflict species: phase out some species, use sterile cultivars, restrict usage to particular geographical areas, agree to eradicate and contain, pay the costs of invasions, and develop options to mitigate risk (including resources to pay for an invasion if it does happen in the future, i.e. insurance). All of these will need some consultative process, and in many cases this might involve self-regulation or accepting risk (Box 6.3). One approach, pioneered in South Africa, is to identify and regulate species that are useful in some contexts, but invasive in others. A specific legislative category was created such that permits can be issued to grow plants providing 'specimens of the species do not spread outside of the land or the area specified in the Notice or permit' (Department of Environmental Affairs 2014).

Box 6.3 *Managing Invasive Ornamental Trees: Conflicting Views and Values in Hawai'i (Curtis C. Daehler)*

In recent decades, a growing awareness that various ornamental plants are or might become invasive environmental weeds has led some environmental advocates in Hawai'i to push for more proactive management of invasive ornamentals and greater oversight of what is planted on public lands or offered for sale to the public. Some invasive trees in

6.6 Challenges to Legislation and Agreements

Figure 1. A Chinese banyan tree (*Ficus microcarpa*) in Wailuku, Maui, Hawaiian Islands. Although there are invasive populations, this individual tree is designated an Exceptional Tree and afforded legal protection. Photo courtesy of Michael B. Thomas.

Hawai'i have a long history of planting in public and private landscaping. One such example is the Chinese banyan (*Ficus microcarpa*) (Box 6.3 Fig. 1), first planted in the early 1900s. The specialised wasp pollinator was then deliberately introduced in the 1930s, and this large tree has since spread into forests where it can grow as a dominating, epiphytic strangler. Seedlings also frequently establish in joints or crevices on buildings, rock walls, bridges, and elevated highways, where their growth can create cracks and structural instability. Management of invasive trees often involves removal of established individuals, along with prohibition of deliberate plantings. In the case of Chinese banyan, fruits are widely dispersed by birds; therefore, seed dispersal from urban plantings can impact neighbouring forests, as well as man-made structures. Management of these impacts would be improved by removing planted specimens. At the same time, several large Chinese banyan trees have been formally recognised as Exceptional Trees (*Hawaii State Legislature Act 105*, passed in 1975), which requires safeguarding them from injury or destruction.

After many hours of discussion and debate among landscapers, non-governmental organisations, and government agencies, a general consensus statement emerged as a starting perspective for any invasive tree management plan:

> Trees are major capital assets in our cities and offer immeasurable social, community, environmental and economic benefits.

> Of course, one might argue that benefits can always be measured, but such measures require making assumptions that not all stakeholders might agree with. The point is to begin the decision-making process by recognising that removing an invasive tree from a landscaped or urban setting will have consequences beyond eliminating that tree as a source of new invasion propagules. The above consensus statement was subsequently published on the first Hawai'i Weed Risk Assessment website (together with further explanation), and this served as an important backdrop to help unite divided stakeholders in recognising a need to carefully consider which trees are being planted and maintained in landscaping. An objective and defensible decision to remove planted invaders and exclude them from future planting is most likely to emerge after understanding the diverse perspectives of stakeholders.

6.6.6 Who Should Pay, Who Can Pay?

One strategy to finance incursion response is to identify and charge those responsible – the polluter pays principle. The manner in which alien species are introduced into a country can be used to define responsibilities if something goes wrong (Hulme *et al.* 2008). For example, importers can be held liable if a deliberately imported species escapes and causes harm. However, insurance companies might be unwilling to insure against the potential impacts of an introduced species as they do not have the information necessary to make the actuarial assessments. This raises the question of whether it is appropriate to permit the introduction of species where an entrepreneur makes the profits, but society bears the costs of a subsequent invasion. This is especially true in cases where the introduction could be regarded as frivolous, and where potential benefits would be far outweighed by the potential risk.

In practice, assigning responsibility is problematic. In particular, it is difficult for exporters, carriers, and developers to ensure that unintentional introductions do not happen; in many cases it would require international law to hold responsible parties to account; and it is often difficult to enforce laws even if the liability is established within a given jurisdiction. Moreover, if the responsibility can be clearly defined in a court of law, the polluter might simply not be able to pay. Ultimately, the state will usually have to intervene if damages are to be minimised.

This has profound implications for incursion response, which, by definition, needs to be proactive. Given the delays in determining liability and tracing the original source, costs will almost always need to be carried initially by a government and be recouped subsequently, although see Box 1.1 for an example where the company responsible for an incursion admitted fault and voluntarily paid for the costs of eradication. Research into different financial and legislative models for dealing with this problem might prove fruitful in dealing with this devilish issue.

6.7 Recommendations

Incursion response managers must be engaged in the legislative process to ensure that regulations can be implemented in practice, and that there is an environment that is supportive of proactive interventions. Separate to the formal legal process, managers and scientists should also engage with stakeholders and facilitate the development of self-regulation (i.e. to develop and maintain environmentally friendly credentials). While this can be very time-consuming, it will often determine ultimately whether coordinated control can be achieved.

Processes	Requirements	Deliverables
Creating proactive legislation	• Identification of potential threats (e.g. using environmental scanning and risk analysis)	• Prohibited and watch lists
Reconciling conflicting policies	• Identification of conflict species • Increased communication between organisations that promote trade and biosecurity organisations • International consultations with industry working groups	• Modernised international regulations and standards

(cont.)

Processes	Requirements	Deliverables
Defragmenting legislation	• Gap analysis of existing legislation • Increased communications between legislative authorities • Enhanced consultation frameworks	• Legislation that addresses biological invasions in one act • Integration of international, national, and regional efforts
Regulating conflict species	• Stakeholder consultations • Risk analyses • Cost–benefit analyses • Identification of alternatives or acceptable risk-mitigating behaviours	• Sound science-base for regulatory decisions • Lists of species permitted under specific conditions of use • Improved trust and compliance in legislative process
Reducing the risks from e-commerce	• Education and awareness programmes targeting e-commerce vendors and buyers • Internet monitoring tools • Standardised international guidelines	• Regulations that target the sale of invasive plants through e-commerce
Financing incursion response	• Identified responsibilities • Legal framework for attributing costs • Actuarial assessments of risk	• Legal procedure to recover control costs • Insurance products covering invasion threat

7 · *Strategies and Actions*

Key questions addressed:

- What should a strategy contain?
- How should stakeholders be identified and engaged?
- How should action items be determined?

Those who are responsible for managing invasive plants are usually confronted with a range of species. As of 2010 South Africa had ∼8750 introduced plant taxa, 660 recorded as naturalised, 198 included in invasive species legislation, but only 64 subject to regular control (Wilson *et al.* 2013). Since then, due to dedicated efforts to detect and document introduced plants, several hundred more species have been recorded as naturalised. How should each species be managed? Which species should be dealt with first? Who should manage the process? The majority of invasive plants are too widespread for any one organisation to manage, or they occupy space such that they proliferate across administrative borders. The invasion might occupy land that is under the care and control of various levels of government, different countries, private landowners, non-profit organisations, and home owners. Invasive plants are a cross-cutting issue affecting public health, national biosecurity, the environment, agriculture, and trading relationships. Decisions have to be made not just in terms of whether there is an overall net benefit from attempting eradication; whether one eradication target is preferable to another; whether resources should be allocated to eradication versus other high-level goals (i.e. prevention, containment, or impact reduction); but also whether resources should be spent on biological invasions rather than other environmental challenges. All of these considerations point to the need for a strategy to guide society's response to biological invasions and to underpin budgetary decisions.

There is a significant body of literature discussing the theory behind strategic thinking. Mintzberg (1978) suggested that the term strategy refers to a deliberate conscious set of guidelines that determines decisions into the future. Mintzberg and Waters (1985) provide examples of eight different types of strategies (planned, entrepreneurial, ideological, umbrella, process, unconnected, consensus, and imposed). Of these eight types, the consensus strategy and the planned strategy are most often adopted to deal with invasive plants. In the consensus strategy, many different actors naturally converge on the same theme and find a common pattern of actions that mutually works. The planned strategy is where leaders at the centre of authority formulate their intentions as precisely as possible and then strive for their implementation – i.e. translation into collective action. Invasive plant strategies are functionally somewhere between consensus and planned strategies, as goals and objectives are formulated by many partners that have converged to address an invasion, but there is usually a single lead organisation.

While legislative instruments can provide frameworks and standards for responding to invasive plants, international, national, and regional strategies are required to support and guide the management of invasive plants (Wittenberg & Cock 2001). For example, the CBD (Section 6.2) has stimulated the development of country and regional strategies that address invasive plants (McGeoch et al. 2010), such as the national Canadian Biodiversity Strategy of 1995, followed by the regional Ontario Biodiversity Strategy of 2011. Strategies against invasive plants are developed and implemented at various spatial and temporal scales, by a wide variety of organisations and people, and are developed often from different starting points (e.g. to manage species, areas, or pathways). There are well over 50 existing international strategies that provide provisions or guidelines on how to manage invasive species (Box 7.1), and many more national and regional strategies. Where national, regional, and local strategies exist in a country, it is important that these are complementary and that linkages between them are explicit.

Box 7.1 *National Strategies for Dealing with Biological Invasions: South Africa as an Example (Brian W. van Wilgen)*

Many countries have ratified the Convention on Biological Diversity (CBD), which requires Member States (among other things) to 'prevent the introduction of, control or eradicate those alien species which threaten ecosystems, habitats or species'. As a first step, most of these countries have developed national strategies for addressing the problem of biological invasions. These strategies typically follow a hierarchical approach, in which priority is given to preventing the introduction of invasive species. The hierarchy also recognises that if a potentially invasive species has been introduced, it should be eradicated if desirable and feasible; that, in the event that resources are not available for eradication, containment measures should be implemented; and that long-term mitigation measures should be implemented where species have become widespread and dominant. Such strategies are usually based on wide consultation between affected stakeholders or arms of government, and they provide direction for high-level policy interventions aimed at achieving defined goals at a national level (Box 7.1 Fig. 1).

In 2013 the South African government commissioned the drafting of a national strategy for dealing with biological invasions, which was completed early in 2014 but is still awaiting parliamentary approval. The strategy adopted a framework to ensure that aspects of the problem at all stages of invasion were addressed (Fig. 1.2), and a number of cross-cutting aspects are addressed in some detail, including improving the legislative and regulatory environment, managing information, building capacity, conducting research, raising awareness and funding, and coordinating activities. The main overarching goals of the strategy are to:

- prevent the introduction of new species that pose a risk of becoming invasive;
- eradicate introduced species where desirable and possible;
- reduce the rate of spread of invasions; and
- reduce the impacts of existing invasions.

Taking a strategic view has already led to the introduction of several new approaches to management. A new government-funded unit has been established to detect and document new invasions, to provide reliable and transparent post-border risk assessments, and to provide the cross-institutional coordination needed to successfully implement national eradication plans (Wilson *et al.* 2013). Steps have been taken to incrementally introduce controls and inspections at airports and harbours. Regulations have been published in which problematic invasive species across all taxa have been listed, and the requirements for control are spelt out. Several other important recommendations in the strategy still have to be implemented, including the adoption and monitoring of a suite of high-level indicators that will be used to assess whether the goals of the strategy are being met.

172 · Strategies and Actions

Figure 1. Cover of South Africa's draft National Strategy for Biological Invasions.

South Africa's strategy differs in several aspects from many other national strategies that have been drawn up for other countries. Given the huge inequalities in wealth and the need for development, the creation of employment and development opportunities, in combination with control efforts, has been the cornerstone of control programmes across the country since 1995 (van Wilgen, Le Maitre, & Cowling 1998). The publication of regulations that govern the management of invasive species has made allowance for species that are both useful in some contexts, and detrimental in others, by creating a suite of categories that deal with species of commercial or ornamental value by issuing permits that include clearly allocated responsibilities for the control of spread by permit-holders. Finally, steps are being taken to make use of biomass (from cleared invasive plants) for the manufacture of furniture and other products, and the generation of energy. Although the effectiveness of these approaches in reducing the spread of invasive species is poorly understood, the schemes are remarkably attractive politically and can generate substantial additional funding.

A number of challenges to the successful implementation of the strategy lie ahead. Although substantial amounts have been spent on alien species control over the past decade, the problem continues to grow (van Wilgen et al. 2012). To become more effective, the country needs to find ways of acceptably prioritising the management interventions so as to achieve optimal outcomes, of raising additional funding (especially from the private sector), of effectively coordinating efforts across affected government departments, and of balancing the imperatives for job creation and development, on the one hand, with effective control, on the other.

A strategy is, of course, of no value if it is not implemented. As such, there needs to be an action plan or a 'to do' list, where it is clear who is going to do what and when, and how it will be paid for. Strategies and action plans are often combined in a single document, but it can be very useful to separate the two to provide focus, for example, an action plan for aquatic or terrestrial invasive plants. Action plans will require much more frequent revision (i.e. adaptive management) and will need to include specific details for funding and responsibilities, whereas a strategy provides an overall direction.

The first part of this chapter will briefly address some of the theory that one might consider in developing an invasive plant strategy, then we focus on some of the common elements found in invasive plant strategies and action plans. Finally, we discuss two of the main steps required to develop an effective strategy: prioritisation and funding.

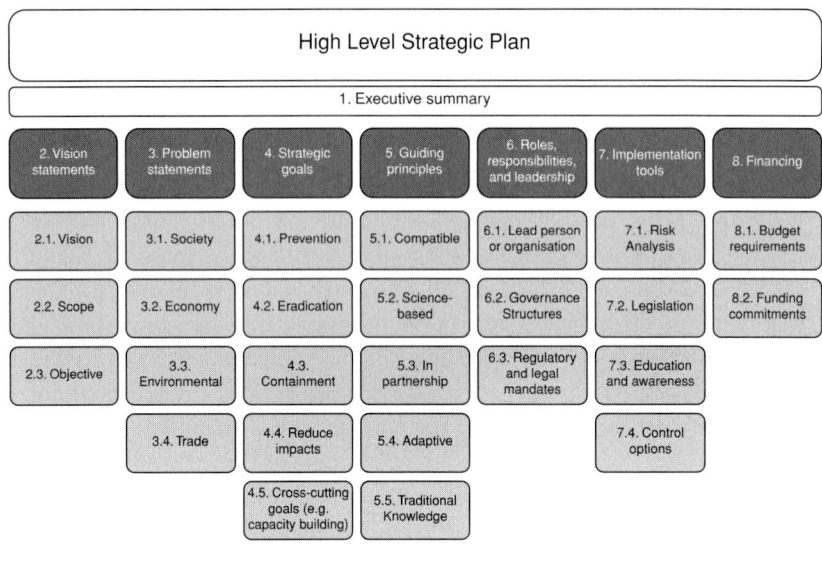

Item	Description
(1) Executive summary	The rationale, key actions, costs, and likely outcomes need to be presented concisely (e.g. in 2–3 pages) with links to the broader detail. This, of course, will likely be the only part that most people read (including the decision makers with the most influence).
(2) Vision statements	One of the first tasks will be to develop vision statements that set the path for the rest of the strategy. Crafting these statements requires careful analysis and discussion on what is the purpose of the strategy and what it will or will not do. Visioning can be described as creating images of the future to serve strategic goals (Shipley 1992). Vision statements should not be treated lightly and it is imperative that all participants are in agreement. The *vision* (2.1) might be as simple as enhancing biosecurity, protecting biodiversity, or preventing the introduction of new species. It should try to encapsulate the strategy's destination and intent. A clear vision will also make it easier to identify potential participants and objectives (Sutherland *et al.* 2011). The *scope* (2.2) of the strategy needs to indicate what is included and excluded. For example, the scope might be all invasive plants or just some select taxa (e.g. terrestrial taxa).

(cont.)

Figure 7.1. A general table of contents for an invasive plant strategy. There are several elements common to most strategies dealing with plant invasions that can be used to guide the development of new strategies. Note that each of the strategic goals (items 4.1–4.5) will require further definition of objectives and strategic actions, sometimes formulated as separate action plans. Invasive plant strategies should receive political sign-off at a high level (e.g. government minister) and form the basis for the development of legislation and regulation.

Item	Description
(3) Problem statements	*Problem statements* document why the invasive plant strategy is needed and help to articulate the forces that spawned it. For example, the problem statement should address the impacts the invasive plant(s) have on (3.1) society (including human health), (3.2) the economy, (3.3) on the environment, and often explicitly on (3.4) trade. These statements help justify the strategy and should provide concrete reasons participants and organisations should invest time and resources.
(4) Strategic goals	The *strategic goals* are statements or targets of what the strategy hopes to achieve based on the vision statements. These should be motivating and empowering. Examples of strategic goals commonly used include (4.1) prevention, (4.2) eradication, (4.3) containment, and (4.4) impact reduction (Fig. 1.2), and should be set explicitly for species, areas, and pathways. There should also be goals related to (4.5) cross-cutting issues, e.g. information management, capacity building, research, awareness raising.
(5) Guiding principles	While the vision and goals might be clearly articulated, it is useful to outline the overall philosophy of a strategy. This can include guiding principles whereby: (5.1) it is explicit how the strategy will interact and be compatible with existing strategies and regulations; (5.2) the approach to decision making is set out (e.g. based on the best available scientific information); (5.3) the intended approach to partnerships; (5.4) the way in which new information will be incorporated into future decisions (e.g. adaptive management); (5.5) and how traditional knowledge will be respected and incorporated into the process.
(6) Roles and responsibilities	Having a well-structured strategy with clearly defined roles and responsibilities is essential. In particular (6.1) it should be clear who the lead person or organisation is; and (6.2) which governance structures are in place or are needed to facilitate the strategy (e.g. Invasive Species Councils). It is critical that the functional roles and responsibilities (and expectations) of the various participants and stakeholders involved in the development and implementation of the strategy are also clearly articulated (e.g. national government; regional government; non-governmental organisations; industry; First Nations). All participants might have a role in some tasks, such as education and awareness, while only a few might have a role in other tasks, such as quarantine inspection programmes. (6.3) The legislative authorities of the participants involved in the strategic plan also need to be (*cont.*)

Figure 7.1. (cont.)

Item	Description
	specifically identified. For example, what are the acts and regulations administered by each participant and what are their legal responsibilities in any response or management programmes? Regulatory and legal mandates need to be well understood as they will dictate who can do what in the event of any incursion. A self-identification process can be used to help identify the strengths and shortcomings of each participant and stakeholder.
(7) Implementation tools	It is also important for a strategy to outline the necessary and sufficient conditions required for implementation. To list a few examples: (7.1) a risk analysis framework is in place; (7.2) there are legislative tools that can ensure compliance; (7.3) there is a strategy for increasing education and awareness; and (7.4) the control options required. This is, however, the point at which the strategy can begin to merge with an action plan.
(8) Financing	For any strategy to succeed it must be clear (8.1) how much is required and for how long the funds are required. (8.2) Funding commitments and expectations of stakeholders over the life of the strategy also need to be clearly identified (ideally as part of outlining the roles and responsibilities).

Figure 7.1. (cont.)

7.1 Important Elements of a Strategy

A strategy is required to initiate or improve upon the capacity to respond, to enhance an existing biosecurity system, or to activate and enhance capacity. A consequence of this is that there are many elements general to all good strategies, including a problem statement, visioning statements (i.e. purpose, scope, objectives), strategic goals, guiding principles, leadership and coordination, and recommendations for funding (Fig. 7.1). For example, a strategy needs to have a clear spatial scope (regional, national, or local), a temporal scope (over the next 5 years, 10 years, or 20 years), and be clear about the focus (on all species, on particular species, or on areas)?

Strategies are multi-dimensional and situational and will therefore vary by industry, government, and regions (Chaffee 1985). Hence, there is no single method, recipe, or algorithm that should be used to design a strategy, but there are some general steps that are required. Cook *et al.* (2014) suggested six key stages: setting the scope, collecting inputs, analysing signals, interpreting the information, determining how to act, and implementing the outcomes (cf. Fig. 7.2).

7.1 Important Elements of a Strategy · 177

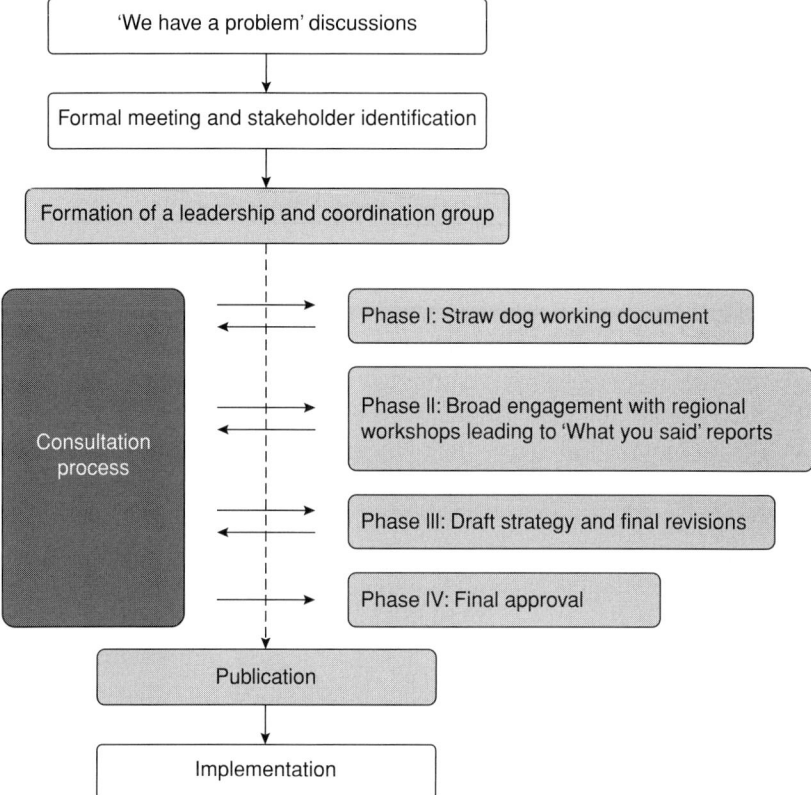

Figure 7.2. Recommended steps in developing an invasive plant strategy. At the heart of the development of any successful strategy is consultation with stakeholders.

What makes some strategies more successful than others? If there is top-down support there is a higher probability that a strategy will be developed, implemented, and funded. For example, the United States of America Executive Order 13112, signed by President Bill Clinton in 1999, stated that federal agencies must not support activities that generally promote invasive species, and mandated the creation of a National Invasive Species Council, a National Invasive Species Management Plan, and an Invasive Species Advisory Committee that includes stakeholders from state government, industry, tribes, academia, agriculture, forestry, recreation, and conservation organisations (National Invasive Species Council 2001). Without the Executive Order in place, it is unlikely that the governance structures and cross-jurisdictional coordination necessary to address biological invasions would have occurred.

But first and foremost, participants and stakeholders need to be fully invested in the process. If there is little willingness to implement a strategy, then it will have little effect. As such, arguably the most important stage in strategy development is stakeholder identification and engagement.

7.2 Stakeholder Identification and Engagement

One of the most resource-intensive tasks in developing a strategy is to ensure that the most appropriate stakeholders are identified and consulted effectively. This process requires considerable resources and time and can take months to years. The general process involves: (1) determining who are the key stakeholders; (2) identifying what the stakeholder needs are and then prioritising these needs; (3) arranging consultations in a manner such that opinions and views can be recorded and analysed in a transparent fashion; and (4) continuing consultations until some consensus or stopping point has been reached or a final 'what we heard' report has been completed. Michaels (2009) referred to this as knowledge brokering, where the intent is to compile information that would otherwise not be acquired.

The first step is to identify the key stakeholders and what is at stake for each. In general terms, stakeholders are those who are affected by the strategy, those who can affect it, and, most importantly, those who are necessary for the strategy to exist (Ackermann & Eden 2011). These will usually include federal, state, territorial, and local government agencies, academic institutions, the scientific community, environmental, agricultural, forestry, and aquatic organisations, trade groups, industry groups, aboriginal groups, landowners, and citizens at large. It is important to engage a broad and diverse set of stakeholders, as a small focused group might heavily influence the strategy in a biased direction. Legitimate and valid stakeholders must be identified early on, both for their contributions to be maximised (Bourne & Walker 2005) and to ensure buy-in. When stakeholders are excluded from developing the strategy, potential allies can become adversaries (Van Driesche & Van Driesche 2004), to the point where a strategy can be delayed or ignored.

In many cases, initialising the collective action required to develop a strategy begins with a national engagement that underpins further regional commitment. The national engagement is often a national forum or summit meeting of some type organised to discuss the need for a strategic response to an incursion. Stakeholders agree that some type of broad cooperation and collaboration is required in a response. From a theoretical perspective, collaboration is deemed necessary, as traditional mechanisms for allocating resources and setting and enforcing rules to effectively secure key public goods have failed (Zadek 2006). Moreover, if you want your stakeholders invested in the process, consultations need to happen early in the process (e.g. in developing the initial drafts). It is preferable to have stakeholders help identify and prioritise strategic goals than asking them to comment on already fixed strategic goals in which they are not invested (Fig. 7.2).

Developing a strategy comes down to a simple question, as presented by Reed *et al.* (2009) – who's in, and why? Stakeholders need to identify clearly who they are representing and whether they have the authority to make decisions or contributions. Stakeholders can then be categorised based on the power or influence they have over the development of the strategy, as well as how the strategy might impact them (Ackermann & Eden 2011; Fig. 7.3). The power or influence a stakeholder brings to the process can be a result of an economic impact (e.g. industry or landowner), personal passion for the issue (e.g. invasive plant council), or political responsibilities to respond (e.g. national or regional government). Stakeholders will also have differing opinions with regard to the severity of impacts associated with an invasion, and bring differing levels of knowledge to the table. For example, an invasive plant might be a deleterious species for one stakeholder group and at the same time a profitable commodity for another. The strategic importance of stakeholders can be assessed by their contribution to the strategy as well as their ability to address an alien plant incursion. According to Mitchell, Agle, and Wood (1997), stakeholders can be latent, dormant, discretionary, demanding, expectant, dominant, dependent, or dangerous, and have the ability to move between these classes. It also should be acknowledged that stakeholders can be observant, conflicted, and reluctant.

A real challenge in developing a strategy is to transparently collect and evaluate input from stakeholders. Some will be very vocal and lead discussions, while others will be more reserved and prefer to provide input through written responses. Often consultations are not able, for one reason or another, to bring all the key stakeholders to the table. As such, there need to be resources for several processes to the engagement (e.g. mechanisms to bring people to meetings or to travel to them, online workbooks and webinars). Interestingly, stakeholders often comment that the key part of the consultation process was the opportunity to meet face-to-face and establish relationships, as opposed to actual strategy development.

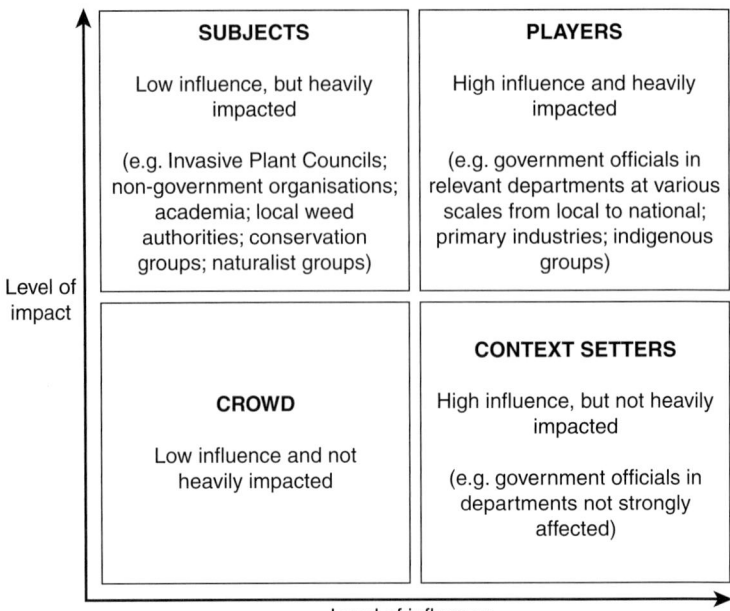

Figure 7.3. A scheme for categorising stakeholders in terms of their level of influence on the strategy and how much the strategy will impact on them. The four quadrants of the grid illustrate the four categories of stakeholder. Most of the people involved will be *subjects* or *players*. However, the *context setters* can influence or even kill the strategy and so it is important to educate them, keep them informed, and seek their input. The *crowd* are potential, rather than actual, stakeholders. The power–interest grid is similar to the stakeholder circle used by Bourne and Walker (2005) to map and visualise stakeholder influence. Both approaches attempt to characterise the attributes of stakeholders based on interest (or impact) and power. Importantly, the position of stakeholders is not fixed and can be influenced. For example, the City of Cape Town's Invasive Species Unit is managing an incursion of the house crow (*Corvus splendens*). Initially there were a few very vocal well-connected people who were advocating that crows should not be controlled due to animal welfare concerns, essentially taking the role of players. Many of the local impoverished communities where the house crows were abundant were the crowd. Through extension work, however, it became clear that the local communities were in fact subjects, as the house crows had direct effects on them – the crows were not wanted. In consequence, the power of the pro-crow lobby declined substantially as it could not be seen to veto the wishes of the general public. In consequence, house crow control could continue (L. Stafford, pers. comm. 2015). Redrawn from Ackermann and Eden (2011) with permission.

Table 7.1. *Tools used to help identify and describe action items during collective decision making*

Name	Description
Pareto voting	The Pareto voting method is a tool that can be used to help prioritise which objectives or action items should be addressed in a strategy. In consultations, you can ask stakeholders to brainstorm and identify action items of interest to the group, and then have them vote or rank items to identify the highest-priority items by applying weights to votes. The method was developed by Vilfredo Pareto (1848–1923), an Italian economist who made theoretical contributions in the analysis of individual choices and devised the 80:20 rule or Pareto Principle (i.e. 20% of items will be chosen by about 80% of stakeholders).
Fish bones	Stakeholders begin by identifying a strategic goal. This is placed at the head of a fish. Possible action items are then listed on the ends of fish bones (branches extending out from an arrow running back from the fish head), and challenges can be identified on the smaller bones leading to each action item. Also referred to as Ishikawa, after a consultant by the name of Dr Kaoru Ishikawa.
Stakeholder roles and responsibilities matrix	The various roles and responsibilities (e.g. prohibit entry, education, and awareness) are listed against spatial information (e.g. not present; widespread in area *x*). Those being consulted are then asked to place sticky notes into the areas of the matrix where they feel they have a role, possibly with different colours for the different stakeholder types (e.g. government, industry). The matrix then becomes a chess board which identifies the various roles and responsibilities of key stakeholders based on the spatial distribution of the invasion.
Horizon scanning	Horizon scanning is a foresight tool used to collect and organise diverse streams of information to identity emerging issues and better understand present issues (Cook *et al.* 2014). In our case, it might be referred to as a type of issue-centred scanning, with the issue being invasive plants

Consultations can take many forms, from face-to-face meetings and regional workshops, to online questionnaires and workbooks. Several regional-based consultations might be necessary when developing a strategy that addresses a large geographic area with many different habitats, climates, and administrative authorities. There are also several commonly used tools available that can be used to help identify and describe action items through collective decision making (Table 7.1). Consideration needs to be given as to who is going to lead meetings or workshops. It is often valuable to hire an independent facilitator to lead the consultations, as well as to summarise the data and prepare a report, since some stakeholders might feel uncomfortable if consultations are led by a government organisation or someone with a similar conflict of interest. Care must be taken in hiring an appropriate facilitator who is unbiased and experienced, yet has some knowledge of invasive plants. There also needs to be some administrative support (e.g. to facilitate invitations, travel, and meeting bookings). Organising and delivering effective consultations is not an easy task.

Some of the most intractable problems in invasion biology come not from the invasive species themselves, but through social conflicts that arise (Woodford et al. in review). As such, it is important to monitor opinions, to see whether conflicts might arise, and to evaluate if particular interventions (e.g. workshops) have been successful in narrowing gaps in perceptions (Novoa et al. 2016).

7.3 Leadership and Ownership

If a strategy is to be successful, it requires a strong, dedicated leader or champion that has the institutional support to ensure the strategy will be taken up. This might be the single most important element in ensuring a strategy is developed and implemented. The champion needs to be able to listen to the views of all stakeholders, but also have the authority to influence stakeholders when necessary to move the strategy forward.

Leadership can come from an individual, but more often it is from a dedicated leadership group. The group must be in it for the long haul as it will take several years just to draft the strategy, never mind implement it. In drafting the strategy a number of aspects should be considered, including provisions for management teams that provide direction and coordination with agreed criteria for achieving management goals and ensuring the appropriate levels of communication take place, as well as financial and resource management planning and information management.

Bryson and Roering (1988) discussed the importance of champions and strategic planning teams in guiding the development of a strategy (they referred to a strategy simply as a deliberate attempt to produce change) through the necessary steps. They argued that successful process champions were usually team leaders, at times good facilitators, and that champions needed to be confident that the strategic planning process would produce desirable outcomes, while also taking into account both expected and unexpected events. In developing a strategy, Bryson and Roering (1988) argued that you require a powerful process sponsor, a process champion, a planning team, an expectation of disruptions and delays, and a willingness to be flexible.

The second essential aspect is that whoever will be responsible for implementing the strategy needs to be invested in the strategy development process and committed to its outcome. There are many examples of strategies that simply gather dust, not because of a lack of leadership, but because they were developed by keen and interested parties who were not going to implement them. So while engaging consultants and contractors to draft sections of a strategy might result in a more erudite and perhaps better crafted strategy, the strategy will not be used unless there is clear ownership of the strategy by whoever is implementing it, and it might be preferable to do such a job 'in-house'.

Many strategies are developed with great energy and spirit, but fail when it is time to implement or move forward to achieve any action items. In such cases it can be important to have an expert working group that is responsible for delivery of specific functions, and that can oversee implementation and evaluation (Cook et al. 2014). It can also be important to have a specific action plan.

7.4 Action Plans

Strategies are often high-level, so it is not always clear what needs to happen in practice. Therefore it can be useful to have a separate action plan that spells out the specifics, i.e. tasks that need to be done to achieve the strategic goals. Action items should be prioritised in terms of how long they will take, which order they need to happen in, and their relative importance. For example, a strategy might identify a need for risk analysis and the associated action plan would then need to include items on how to develop risk assessment and risk management tools. But without capacity in place to do this, nothing will happen, and so the first action item might be to recruit, contract, or train staff to develop a risk assessment framework. Another common strategic goal is to develop an early detection and rapid response system. In the associated action plan, there could be action items to develop active and passive surveillance networks,

Table 7.2. Template for an eradication action plan. Some generic goals and long-term indicators are provided. Other goals and indicators, plus annual objectives and measures of success, will need to be determined according to the specific circumstances of the incursion response. Adapted from www.cdc.gov/prc/pdf/2009foa-appendix3.doc with permission.

Project title: Eradication of [targeted species]

Project period goal:
Over the period [*start date to end date*] this programme will:
(1) Delimit the extent of the incursion via surveillance and tracing activities
(2) Prevent further spread and extirpate individual populations by controlling plants before they can reproduce
(3) Identify and commission research required to improve management effectiveness and efficiency
(4) Maintain stakeholder consultation

Long-term impact or outcome: Eradication is approached, if not achieved, within the specified period.

Long-term indicators:
(1) Given continued surveillance, the amount of newly detected invaded area decreases; delimitation measure in eradograph approaches zero
(2) Plant density decreases, with few or no plants escaping control; extirpation measure in eradograph trends negatively
(3) Lack of information does not hinder operations
(4) Stakeholders remain engaged and supportive of the programme

Annual objectives	Measure(s) of success	Activities	Person responsible	Completed? (Y/N)	Completion date
Precise, time-based, and measurable actions that support the completion of project period goals (e.g. 1–4).	Standards for each of the project period goals to measure progress in achieving an annual objective. These should be numeric or capture clearly observable behaviour.	Key events or actions implemented to achieve specific annual objectives			

provide support to herbarium staff for identification, and build capacity to implement emergency control measures. Action items need to be mutually agreed upon and come from the stakeholder consultation process, providing specific activities for stakeholders to rally around and implement change. Action items, like strategic goals, can only be achieved if there are available resources and budget, so it is important to identify any existing resources and programmes that could be used to achieve them. Ownership needs to be assigned to each action item so it is clear who is responsible for it and to ensure it gets done. In some cases the completion of one action item is critical as it leads to subsequent action items, so timelines become important.

Action planning needs to occur at multiple levels. For example, the higher-level planning considered above must be complemented by additional planning at the eradication programme level (Table 7.2), should this be the preferred management goal for a specific incursion. Here, key actions associated with surveillance, control, and communication will be identified, along with timelines for each. Such lower-level planning might be conceived as a nested approach, whereby annual action plans reside within plans spanning longer time frames. Planning for both time frames needs to be flexible according to contingencies that might arise during the course of implementation. For instance, discovery of substantial new invasion foci could significantly extend the time frame required for eradication, whereas technical developments, such as new methods for achieving rapid depletion of seed banks, could have the reverse effect. And of course, major deviations from the level of funding indicated at programme commencement might necessitate a revisiting of management goals (see Chapter 5).

7.5 Funding and Prioritisation

When stakeholders are asked to identify the major challenges facing the management of plant invasions, funding is consistently among the top limiting factors. Funding is required to develop the strategy, its strategic goals, and any associated action plans. However, invasive plant management has rarely, if ever, attracted the level of funding that the management of invasive diseases and animals does (Panetta 2009). For instance, the ongoing eradication programme targeting fire ants (*Solenopsis invicta*) in south-eastern Queensland, Australia has cost more than AUD 300 million since its inception in 2001. This stands in contrast to major eradication programmes against branched broomrape (*Orobanche ramosa* subsp. *mutelii*) which commenced in 2000 and chromolaena (*Chromolaena odorata*) which commenced in 1994; both programmes were abandoned in recent times and entailed expenditure of approximately AUD 45 million and AUD 10 million, respectively. While there were some technical issues that were exceedingly difficult to overcome in the case of branched broomrape (e.g. very high seed persistence), there is good reason to believe that chromolaena might have been eradicated had more money been invested early on – annual expenditure in the first ten years of this programme averaged only AUD 185 000 (Panetta 2015).

Therefore, prioritisation is vital at all levels of the management process, from the allocation of resources between the overarching strategic goals to between specific action items. For example, decisions will be required as to how much should be spent on prevention versus eradication, which species to tackle first, and whether to tackle one population before another. Without prioritisation there is often a bias towards reactive management as it is seen as less risky (Section 4.1; Finnoff *et al.* 2007).

This is one example where, in bioeconomic terms, the best approach to take is one of benefit maximisation. This procedure identifies the optimal outcomes for a fixed budget, and helps decision makers to select the species that should be targeted first, and the ones for which eradication programmes cannot be undertaken without a larger budget (Parkes 2006). Often there are multiple competing factors that need to be taken into account, both environmental and socio-economic (and sometimes political), as well as multiple views and perspectives. There are several decision support tools that can be used to work through the problem. For example, Roura-Pascual *et al.* (2009) used the DPSIR (driving forces–pressure–state–impacts–responses) framework and the analytic hierarchy process (AHP) to guide prioritisation for managing plants in the Cape Floristic Region of South Africa. Hohmann *et al.* (2013) used a similar technique in North Carolina, USA.

To some extent the problem can be reverse-engineered by determining actions based on the resources that are (or are likely to become) available. Resources might be entirely consumed by dealing with threats that have previously materialised elsewhere, i.e. where there is strong evidence that the invasive plant can cause serious damage. In this case the emphasis will be on the likelihood of damage being realised (again in the jurisdiction concerned) and the potential speed at which it might develop. An alternative, which will address the problem of species that have no history of invasiveness and impact, is to deal with all incursions below a critical

Table 7.3. *Strategic goals adopted by the Invasive Species Council of Manitoba, Canada in 2014, with relative allocation of resources between each*

Priority	Time and resources (%)	Goals
Goal #1	30	A strong functional council
Goal #2	25	Provide leadership by being a provincial coordinating body
Goal #3	20	Foster innovative education and awareness
Goal #4	15	Prevent new introductions through early detection and rapid response (EDRR)
Goal #5	10	Maintain a provincial invasive species database (e.g. using the EDDMaps application, Section 3.3)

size or stage (Harris & Timmins 2009) within constraints on budget and resource availability. Similar to the last option, it is also possible to select all targets that are likely to provide a given return on investment (regardless of the absolute cost). Fundamentally, for any of these prioritisations to work, it is essential to put some constraints upon the system. Within these constraints, strategic goals can then be prioritised or ranked to highlight what needs to be done first and which items are more important than others (Table 7.3).

In many cases strategies fail simply because of a lack of long-term funding, and this is particularly the case for incursion response, where eradication generally requires long-term funding and commitment over typically ten years or more (Simberloff 2003; Panetta 2007). Given the time scales over which a strategy might be needed, arrangements need to be in place to ensure continuity of funding and infrastructure. Of particular relevance here is the need to develop 'motivating cases' for the allocation of resources to individual programmes.

Funding for invasive plant strategy development can come from a variety of sources (e.g. federal, provincial, and local governments, corporations, private funders, philanthropies), but the costs of strategy development are usually much, much lower than the costs of implementation, and development will usually be a relatively short, intense engagement. Despite the significant economic and environmental impacts of invasions, long-term sustained funding remains a major challenge for strategy implementation. During the life of the strategy, governments will come and go, national and regional priorities will change, people will retire, champions will move on, and the 'invasive plant of the day' might slip off agendas and budgets. There is also a jurisdictional issue – is it a national responsibility or a local responsibility? A key component in ensuring funding is to 'make the case' in a way that will catch the attention and imagination. This will likely need both some headline economic figures and some specific case studies to make the issue more tangible (Box 7.2).

> Box 7.2 *Costing Invasions in the UK (Richard H. Shaw)*
>
> It is clear that in many countries decision makers are more influenced by economics than ecology, as evidenced by the push to value ecosystem services as a way of raising their status and therefore level of protection. This is particularly so for cross-cutting issues such as invasive species which are everybody's problem but nobody's responsibility. In Great Britain it was recognised that there was a lack of information on the economic cost of invasive species to the country, so CABI were commissioned to deliver a report on the subject by the Department for Environment, Food & Rural Affairs (Defra), the Scottish Government, and the Welsh Assembly in an attempt to get some solid figures that would aid prioritisation.
>
> We considered the cost of invasive non-native species (INNS) across 12 different sectors, namely: agriculture, horticulture, forestry, quarantine and surveillance, aquaculture, tourism and recreation, construction development and infrastructure, transport, utilities, research, biodiversity and conservation, and human health. Information on non-native species – including weeds, vertebrates, plant pathogens, and invertebrates – was to be gathered from scientific and grey literature, as well as the internet. Additionally, questionnaires and interviews were to be conducted with key organisations, including policy makers, land owners, and managers in order to calculate further estimates.
>
> A total of 730 questionnaire consultations were conducted, 250 telephone calls made, and more than 650 references of relevance gathered from scientific and grey literature. This included information on 500 non-native species.
>
> As a result of this extensive study, the cost of INNS was estimated to be around GBP 1.7 billion per year, specifically GBP 1.29 billion in England, GBP 0.25 billion in Scotland and GBP 0.13 billion in Wales. Rabbits and Japanese knotweed were identified as being responsible for the majority of the costs incurred (e.g. Box 7.2 Fig. 1), followed

Figure 1. Japanese knotweed pushing over a wall in Reading, UK. This species is estimated to cost the British economy GBP 165 million per year. Photograph courtesy of Richard H. Shaw.

by wild oat and the brown rat. In addition to this, the report highlighted how the costs of control increase dramatically as invasions progress. Using case studies, early identification and intervention aimed at eradicating invasions before they become widespread were highlighted as of particular economic benefit.

The study was no easy task and necessitated a species-based approach rather than a sectoral one, and was heavily reliant on expert opinion and the grey literature. One of the biggest challenges was bringing together scientists and economists and gaining acceptance of the data and their transformation to pounds sterling. Economists can't believe that the financial data were almost never gathered in the first place, not least to justify an intervention, and the ecologists couldn't believe it was acceptable to extrapolate figures from so few data. Nonetheless, all realised that in most cases the lack of direct economic data did not actually prevent cost estimates from being made.

Another lesson learned was that what is possibly the most in-depth review of INNS costs any country has carried out should have been published in a peer-reviewed journal, but instead was published online without a digital object identifier (doi), which precluded publication and limited its impact in academic circles. Nonetheless, a follow-up project three years later reviewed the impact of the report and revealed that the results (in particular, that the overall cost of INNS was estimated to be GBP 1.7 billion) have reached a very wide audience, including academics, Members of Parliament, and the general public. The report is cited well in the literature, and a high proportion of citations are in high-impact journals. According to Google Scholar, the report had been cited 97 times in journal articles as of June 2016 and it has appeared in well over 100 media articles, including articles published in France, Malta, Mexico, Russia, and the USA.

However, the most important goal of the work was to influence policy makers; follow-up survey efforts revealed that it had been well received and well used. When the report was first published, Jane Davidson, Minister for Environment, Sustainability and Housing in the Welsh Assembly Government, acknowledged that: 'This report will help us to prioritize and target where actions can have the most impact, and will assist us in prioritizing our resources for action in the future.' The economic impact of INNS was quoted at the House of Commons Environmental Audit Commission as recently as 12 March 2014 by Parliamentary Under-Secretary of State for

Natural Environment and Science, Lord de Mauley. It was presented at the Environment Audit Committee Invasive Species Enquiry in January 2014, and included in *Biodiversity 2020: A Strategy for England's Wildlife and Ecosystem Services Report* available from the UK Government website. The Committee also called upon CABI to provide an expert witness to the inquiry.

Individual feedback from key stakeholders and policy support personnel confirms impact with reference to the report's importance in awareness raising, providing evidence for action, especially with respect to early intervention, and acting as leverage for funding: 'The report has already been enormously influential; it's difficult for me to overestimate its importance' (Dr Niall Moore of the GB Non-native Species Secretariat).

In summary, producing a thorough and defensible economic assessment of the impacts of invasive species for a country or region is a very challenging but extremely useful activity. It makes the public and its elected representatives pay attention to a cross-cutting issue that is under-represented in decision making the world over.

For a review of the process and impact of the assessment, see Shaw *et al.* (2014). For the actual assessment, see Williams *et al.* (2010) at: www.nonnativespecies.org/downloadDocument.cfm?id=487

7.6 Recommendations

Strategies should contain some important generic elements, and follow a general process, but also be flexible. Stakeholders need to be identified and engaged; there are various tools to achieve this. A champion has to be identified and given the mandate to ensure the strategy can be developed. The organisation that will be responsible for implementing the strategy must have ownership of the strategy and be provided with sufficient resources for effective implementation and monitoring. An action plan, giving details of how things will happen and when, is essential. Both the strategy and action plan need to have clearly set-out, prioritised goals, with details of how both are to be funded. To this end, the reason for management must be clearly and forcibly articulated.

Processes	Requirements	Deliverables
Strategy development	• Clear scope (scale, time frame, taxonomic or functional groups considered) • Problem statement • Vision statement • Prioritisation of broad activities relating to the management of plant incursions • Clearly defined roles and responsibilities for participants and stakeholders	• Documented process of strategy development and consultation • Goals that are Specific, Measurable, Attainable, Realistic and Time-bound (SMART) • Estimated requirements for funding • Workshop reports • Stakeholder consensus
Stakeholder identification and consultation	• Stakeholder identification exercise • Workshops run by dedicated independent facilitators • Options for written or web-based input	• List of stakeholders and power relationships • Records of consultations
Determining action items	• Prioritisation exercises • Management goals and timelines • Budget and timelines	• Action plan • Plan for monitoring and evaluation

8 · *Implementation*

Key questions addressed:

- How should an incursion response be organised?
- What types of partnerships are needed?
- What can be done to increase preparedness?
- How can communication and outreach be enhanced?

Many strategies and plans are developed in great spirit and with great enthusiasm, but fail when it comes time to take action. There are unfortunately far too many beautifully produced and insightful strategies gathering dust on bookshelves. Successful implementation will require risk analysis, legislation and regulations, science, education, and outreach, and international cooperation. Subject expert working groups are often formed to deliver on specific functions such as risk analysis. In this chapter we focus on some of the factors that are required for successful implementation and in particular look at examples where successful partnerships have been developed.

8.1 Organisational Structure and Response Planning

Incursion response is rarely a linear process of detection, identification, risk analysis, selection of a management goal, implementation, and evaluation (Fig. 8.1). As soon as an incursion is detected, initial efforts to respond should be made pending decisions related to whether coordinated control would be supported and the estimates of efficacy of various management options (Chapter 4). This might simply take the form of attempting to prevent further spread, but will certainly involve efforts to define the spatial extent of the problem (Chapter 3). In cases where there is little information on the targeted species, the failure of management activities will require adaptive management, sometimes including

Implementation

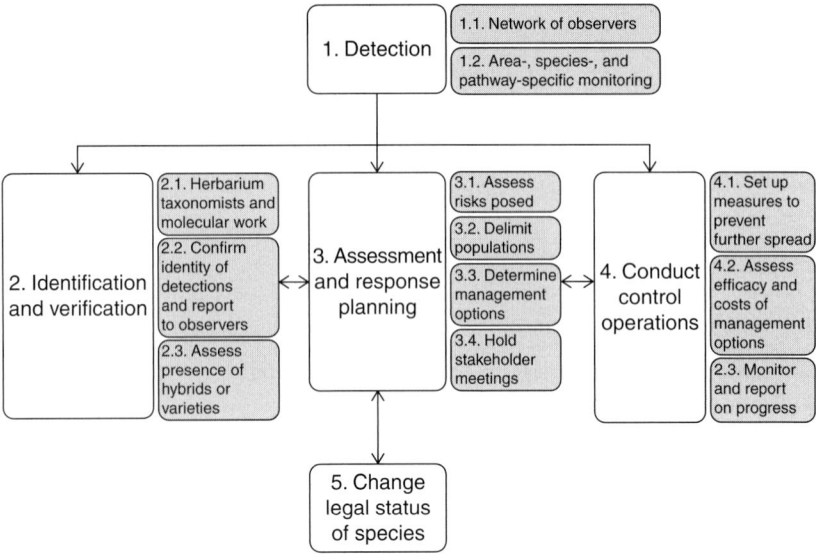

Figure 8.1. Components of a detection and response programme. Based on the scheme developed for South African National Biodiversity Institute's Invasive Species Programme (Wilson et al. 2013).

feedback into an assessment of management options (Chapter 5). If only a few plants are involved, an eradication programme might be relatively straightforward. However, formal efforts at detection and delimitation usually turn up new populations, often distributed across administrative boundaries. As such, an effective response will require substantial organisation, often involving multiple stakeholders (Chapter 6), requiring collaborative partnerships between various levels of government, different countries, private landowners, and non-profit organisations. It will also often require specific research questions to be addressed.

In terms of response planning, it must be clear that the general requirements for eradication (IPPC-FAO 2006) are being satisfied, including the key steps of establishing a management team and a business plan, as well as documentation of the surveillance, containment, and treatment activities required to achieve the management goal (Table 8.1).

8.1.1 Leadership

As for developing a strategy (Section 7.3), perhaps the most important aspect for ensuring that implementation is effective is strong leadership. Leadership might come from an individual but more often it is in the

Table 8.1. *General requirements for pest eradication programmes and an example of its implementation. The requirements are as per the International Standards for Phytosanitary Measures (IPPC-FAO 2003), and the example of implementation is from the Australian National Siam weed (*Chromolaena odorata*) eradication programme. Reproduced from Biosecurity Queensland (2009) with permission.*

Requirement	Implementation in the Australian national Siam weed eradication programme
Establishment of a management team	Tropical Weeds Management Committee and a project coordinator
Establishment of a plan	National Siam Weed Business Plan 2009–2012
Surveillance	Extended survey by helicopter and boat. Ground survey of 200 m buffer around each recorded infestation. Passive surveillance by implementing the Communication Strategy.
Containment	Formal and informal weed hygiene activities are conducted. Delimitation surveys by helicopter.
Treatment	Most effective herbicide application methods selected after scientific research. Fire as a management tool also being implemented and researched in the dry tropics.
Verification of pest eradication	Seed longevity research has established seven years of nil recruitment before a site can be classified as eradicated. Database and biological performance indicators provide information of programme effectiveness.
Documentation	Records are kept throughout the programme (e.g. Tropical Weeds Database) and data represented via annual reports.
Programme review	Thorough independent review in 2003 and 2008. Updated cost–benefit analysis completed in 2008.
Declaration of eradication	Not finalised

form of a dedicated leadership group. The group must be invested and resourced for the long haul. Provisions should also be considered for leadership teams that can provide direction and coordination with agreed criteria for achieving management goals and ensuring that appropriate levels of communication take place, as well as overseeing activities such as financial and resource management planning and information management.

Historically, many incursion responses have been led by people for whom the work is not a core part of their job; this has often meant that for longer-term projects the project is completely dependent on a single person. Strategies have ultimately failed due to the champion changing jobs or otherwise being diverted towards different responsibilities. Ensuring the continuity of a programme beyond the working life of the person who initiates the work is a major requirement (and challenge), and requires that efforts are appropriately documented and evaluated.

8.1.2 Addressing Knowledge Gaps

A dedicated research effort might also be required to provide a basis for a more effective response. Prevention (or minimisation) of reproduction is critical to the success of both eradication and containment efforts, but time to reproduction might not be known for growth under local conditions. Setting the site visit frequency according to information obtained elsewhere can result in substantial seed production if growth is faster than recorded previously (Panetta 2015). Control methods are often initially 'best guesses', providing considerable scope for refinement, and it might similarly be possible to improve surveillance methods and mapping technologies. Such issues are often best addressed by issuing contracts for specific technical pieces of work. Providing there is capacity available, contract research is usually the best way to ensure a result is produced in a given time frame, although, of course, it can be costly.

An alternative, longer-term way of bridging knowledge gaps is through funding postgraduate students or university researchers. In these cases there are significant risks that the work will not happen to the standard required, and in some cases a student project might simply fall apart. For this approach to work, there needs to be strong, ongoing partnerships built with universities and specific academics. Once a relationship of trust is built up over time, it is much easier for the applied research required to be focused on producing a useful answer, while still meeting the standard required for publication. There are various ways to bridge what has been dubbed the knowing–doing gap (Esler et al. 2010; Shaw, Wilson, & Richardson 2010), but in all cases we feel it is important that there is time available for academics and managers to interact on a less formal level.

8.1.3 Monitoring and Evaluation

As discussed in Chapter 5, there are several relatively simple metrics that can be used to measure the effectiveness of a management response, both at a population level and at a programme level. The most appropriate metrics to use can vary depending on whether the overall goal is eradication, containment, or asset protection, but they all require some level of data collection, analysis, and interpretation. There can also be value in separating control from evaluation. Such evaluations will often need to wait until it is easy to differentiate controlled plants from ones missed, or at an interval sufficient to pick up any regrowth of treated individuals. By incentivising both the people doing the evaluation and the control teams (e.g. in terms of finding plants), it can be possible to increase effectiveness.

Once the data are collected, there needs to be a process whereby progress is assessed and results feed back into on-ground operations (i.e. indicators are used so that management can be adaptive). Which teams are efficient? Which control methods are successful? What areas require follow-up work? Coordinators need to set up regular meetings where the data collected can be interrogated and the experiences reviewed. A set process of progress review that potentially incorporates population and decision support models is essential. This is where coordinators can ensure that researchers and managers are interacting regularly (i.e. the knowing–doing gap is bridged).

Monitoring and evaluation also provide an important opportunity for continued consultation with stakeholders. In addition to providing evidence of progress, regular meetings with stakeholders will help to ensure continuity in the face of inevitable staff turnover over the course of long eradication campaigns as well as helping to keep stakeholders informed and invested. Such consultations are also important to ensure that stakeholders continue to feel invested in control programmes and that they can see how their contributions (or lack of them) have affected management.

An additional role of monitoring and evaluation is to provide feedback to the funders. Ideally, progress reports to funders should be based on metrics that reflect progress towards achieving management goals. However, this is not always the case. For example, the Working for Water Programme in South Africa has been enormously successful in creating jobs based on the need to manage invasive species (in particularly woody species that use water). However, the lack of monitoring and evaluation in relation to ecologically relevant management goals means that it has been very hard to gauge progress either nationally or locally. As such, Working for Water has been based on a 'strategy of hope' (van Wilgen et al. 2012). Similarly, eradication campaigns have been severely compromised by political pressures to work on highly visible species rather than those that are appropriate targets (Wilson et al. 2013). Making sure that reporting requirements do not compromise the likelihood of success on the ground requires strong leadership to act as a buffer, as well as appropriate bureaucratic structures.

The final, and perhaps most important, aspect of monitoring and evaluation is the need to determine when the management approach needs to change. Such change might be either quantitative or qualitative. Quantitative change could involve adjustment in how and where management effort is applied: intense search-and-destroy operations might be required in particular areas missed between surveys, while less effort might be required in some other areas. Abandonment of eradication as a management goal and switching to an alternative strategy (e.g. containment or maintenance control) represent qualitative changes. It is worth noting that containment might be a valuable stop-gap measure until other management tools (e.g. classical biological control) are developed.

A recurring feature is that management activities might be compromised due to the resistance of particular stakeholders to the programme. There are various ways in which such issues can be resolved. In particular, it is important to try to facilitate interaction and seek compromise, although if this fails it can come down to two options: first, accept that a goal is not achievable due to the intransigence of a particular stakeholder; or second, enforce compliance (Novoa et al. 2016).

8.1.4 Compliance

Who should implement control measures? Experience has shown that timely control (i.e. killing plants before they are able to reproduce) is most likely to be achieved when carried out in a coordinated manner by capable and committed land managers. However, in many cases government staff or their contractors will have the most relevant experience, be suitably resourced, and have the mandate to carry out the control across the whole range of an incursion. Various cost-sharing initiatives have been developed and notices of compliance can be issued to whomever is ultimately legally responsible. But unless there is a clear authority with experience in enforcement (e.g. national plant protection organisations), access to private property can be a major, if not insurmountable, problem in many incursion responses.

A more conciliatory approach is to create incentives to comply (e.g. compensation payments, tax credits, or cost-sharing agreements). Native plants can be offered in direct exchanges for the removal of aliens, and the costs of removal carried by society. However, in cases where people hold substantially different values, it can be difficult to convince people to comply voluntarily. Many landowners want to manage their land as they wish, and any government legislation or regulation is seen as excessive. In addition, as in so many conservation initiatives, a few fanatical individuals can prevent the achievement of agreed societal goals. As a last resort, there needs to be room to enforce both local ordinances and national regulations. With regard to invasive plants adversely impacting the environment, however, few cases have gone to court, but we suspect it will become an increasing feature of management. However, enforcing compliance requires a whole additional set of new skills (e.g. lawyers), and the immediate benefits to a particular control programme might come at the cost of societal good will. It is also preferable if the arguments for control do not come solely from the government. Local champions can be far more effective and much less intimidating than if the message comes from an external position of authority.

There are other cases where landowners are keen to comply, but simply don't have the resources to do so. Legislative efforts can introduce costs that the landowner might not be able to meet, might prohibit farmers from selling their crops in the most lucrative market, or might significantly reduce the value of the land. As such, while laws and regulations should be enforced, there will often need to be some allowance made to ensure that the management goals are achieved. This will require partnerships.

8.2 Partnerships

Plant invasions cross administrative boundaries and hence management responses also need partnerships and alliances. This requires a level of organisation that a priori often does not exist. Invasive plants can impact a wide, diverse group of stakeholders and often these incursions have unexpected impacts on unsuspecting stakeholders. Even within an agency or organisation, responding to an invasive plant incursion can involve numerous departments, which all have some responsibility for invasive plants. Developing partnerships to detect and manage invasive plants is therefore crucial (Lindgren 2002).

According to Himmelman (1992), a collaborative partnership can be seen as an alliance among people and organisations from multiple sectors working together to achieve a common purpose. Collaborative partnerships can take many forms, including task forces, projects, coalitions, consortia, and grassroot initiatives, and are used to address invasive plant incursions across broad spatial scales or within local communities. The requirement for partnerships is often driven by funding agencies. Some projects are funded based on the sheer number of partnerships – the more partners the better the chance of being funded. And national governments and biosecurity organisations also actively pursue collaborative partnerships to manage invasive plants.

The following sections (and Boxes 8.1 and 8.2) give examples of how strategies and action plans have been implemented against widespread invasions, both nationally and internationally. These examples illustrate the development and activities of partnerships tasked with managing high-profile weeds, but partnerships are also essential when managing new incursions. For example, 13 organisations contributed to the Southern California Caulerpa Action Team, and this partnership is cited as one of the reasons why *Caulerpa taxifolia* could be eradicated (Anderson 2005). Insights gained from the use of partnerships in the management of widespread invasions will likely also apply to incursions.

Box 8.1 *What is a Cooperative Weed Management Area? (Al Tasker)*

Cooperative Weed Management Areas, or CWMAs, are partnerships within a specific local area comprising federal, state, and local government agencies, tribes, individuals, and other interested groups that manage noxious weeds or invasive species. The initial CWMA was formed in 1992, by the Greater Yellowstone Coordinating Committee to establish a unified strategy for invasive plant management after a large fire dramatically altered their landscape. The success of this first CWMA (which still exists in 2015 – www.fedgycc.org) encouraged others to organise similar groups that allow collaboration and cooperation among agencies. CWMAs might have different names in different regions; for example, Partnerships for Regional Invasive Species Management (PRISMs); Cooperative Invasive Species Management Areas (CISMAs); or Invasive Species Teams or Partnerships. They can be organised in a variety of ways, but they share six basic characteristics:

(1) They operate within a defined area, distinguished by a common geography, weed or pest problem, community, climate, political boundary, or land use.
(2) They involve a broad cross-section of landowners and natural resource managers within the CWMA boundaries.
(3) They are governed by a steering committee.
(4) They have a long-term commitment to cooperation, usually through a formal agreement among partners.

(5) They have a comprehensive plan that addresses the management of invasive species within their boundaries.
(6) They facilitate cooperation and coordination across jurisdictional boundaries.

CWMAs have become an effective way to apply for, receive, and prioritise resources for invasive species management. While money might be the initial driving force for organising a CWMA, other lasting benefits include increased local communication, cooperation, and trust. These collaborations among a broad range of stakeholders have over the last 10–15 years been at the hub of many successful attempts to prevent, mitigate, or eradicate invasive species from given locales. As a result, CWMAs now occur coast to coast in North America (www.naisn.org/cwmamap).

Much of the above information is quoted from the CWMA Cookbook, which can be found on the web in various forms. www.mipn.org/CWMACookbook2011reduced.pdf

---

Box 8.2 *The European and Mediterranean Plant Protection Organization: Coordinating the Response to Invasive Plants Across Borders (Sarah Brunel)*

The European and Mediterranean Plant Protection Organization (EPPO) is an intergovernmental organisation responsible for cooperation in plant protection among its 50 member countries (www.eppo.int). The organisation began in 1951 with 15 member countries and has since expanded to 50 countries that exhibit a willingness to collaborate on plant health matters. Since the 2000s, EPPO has extended its range of activities to plants as pests (i.e. invasive plants), including those having deleterious impacts on the environment. The EPPO Secretariat comprises 14 staff members. It is based in Paris and provides recommendations and guidance to its member countries on plant health. These are provided under the form of standards, lists of species, databases, or through the organisation of conferences, workshops, and training courses. As EPPO works at the regional level, it strongly relies on the participation and expertise of representatives from its member countries. EPPO documents are produced by dedicated panels and then revised by various bodies before being approved by consensus by its member countries.

EPPO has helped develop networks of experts and early detection and rapid response plans for invasive plants, as well as for classical pests (e.g. bacteria, insects, nematodes, and viruses). EPPO publishes a free monthly reporting service (www.eppo.int/PUBLICATIONS/reporting/reporting_service.htm) providing short notes on events of phytosanitary concern, including invasive plants. New geographical records, management measures, pests intercepted in trade, and changes in the legislation are gathered from the scientific literature, as well as from reports from plant protection organisations, and are disseminated through a mailing list of more than 5000 people. Browsing the scientific literature and being in close contact with its network of experts allows EPPO to find material to fulfil one of its missions: identifying emerging invasive plants for the Euro-Mediterranean area. An 'Alert List' captures emerging invasive plants. These species are assessed once per year against the EPPO prioritisation process for invasive plants by a dedicated panel (EPPO 2012). Such assessments allow the species that represent the highest risk to be determined and prioritised for an EPPO Pest Risk Analysis (PRA).

A PRA in the EPPO framework is defined as 'the process of evaluating biological or other scientific and economic evidence to determine whether an organism is a pest, whether it should be regulated, and the strength of any phytosanitary measures to be taken against it' (IPPC-FAO 2013a). A PRA represents a justification for the imposition of preventive measures on international trade. To conduct PRAs for classical pests, as well as for invasive plants, EPPO has developed over many years and through research projects a decision support scheme for quarantine pests (EPPO DSS) (EPPO 2011). This scheme is consistent with international requirements, in particular with the International Standard for Phytosanitary Measures number 11 of the International Plant Protection Convention (IPPC-FAO 2013b). The elaboration of an EPPO PRA is done through an Expert Working Group of five to eight experts knowledgeable about the species under assessment, the crops or habitats at risk, modelling techniques, or the EPPO DSS. Working with a draft PRA previously prepared, and on a basis of almost all the available literature on the species considered, over four days the Expert Working Group produces a comprehensive document assessing the probabilities of entry, establishment, and spread of the species, and the magnitude of its impacts on crops, on the environment, and its potential social impacts (e.g. on human or animal health). The latest available publications are taken into account, as well as cutting-edge climate and land-use modelling techniques, ensuring that science soundly informs decision making. Preventive management measures are also proposed in the framework of the PRA if the species qualifies as a quarantine pest.

The completed PRA undergoes a series of reviews before being approved by the 50 member countries. If the PRA determines that the species represents a risk, the species is recommended for regulation to EPPO countries and placed on EPPO lists. It is then the responsibility of EPPO countries to include the species in their regulations.

For invasive plants recommended for regulation by EPPO, guidance on monitoring, eradication, and containment are also published as standards (http://archives.eppo.int/EPPOStandards/regulatorysystems.htm) for countries to be able to prepare to undertake rapid action in case of an outbreak. This is indeed the ultimate responsibility of countries: to undertake the field surveillance, eradication, and containment actions.

Providing recommendations to prevent incursions of invasive species at the scale of 50 member countries having different climatic, agricultural, economic, and social realities also involves giving adequate training on the tools developed. Training courses are therefore frequently organised for EPPO countries on the EPPO prioritisation process for invasive plants and on the EPPO DSS (www.eppo.int/INVASIVE_PLANTS/ias_plants.htm). Beyond sharing knowledge, such events are important for countries to exchange experiences and to strengthen the regional network of experts.

Recently, European and Mediterranean countries have realised the importance of communication and awareness-raising campaigns when dealing with incursions of invasive species (see Brunel (2014) and the presentations of the EPPO/CoE/IUCN Workshop 'How to communicate on pests and invasive alien plants' at http://archives.eppo.int/MEETINGS/2013_conferences/communication_pt.htm). Training courses for civil servants, farmers, administratiors, in addition to information to the media and citizen science projects to monitor invasive species, are increasingly seen as necessary steps for an efficient response to an incursion. As a regional plant protection organisation, EPPO plans to use *Parthenium hysterophorus* as a case study for communication actions in the region. In the EPPO region this noxious weed is so far present only in Israel, but could potentially threaten the whole Mediterranean basin and the warmer temperate European areas.

8.2.1 Purple Loosestrife (*Lythrum salicaria*)

Purple loosestrife is an invasive wetland plant causing deleterious environmental and economic impacts across North America (Pimentel *et al.* 2001; Lindgren 2003). It is believed to have been introduced into North America in the early 1800s in the ballast of ships arriving from Europe, but subsequently intentionally spread through beautification projects which saw waterfowl hunters spreading seed through wetlands and marshes in hopes of adding colour. Purple loosestrife is now present in every Canadian province. It has the ability to invade natural areas and replace native plant species required by wildlife for food, shelter, and breeding areas.

A number of agencies realised that no single agency or group was in a position to respond or fund a management programme against purple loosestrife. Collaboration was necessary. Hence, a multi-jurisdictional partnership was formed, comprising local community groups, provincial and federal agencies, and non-profit groups. The group was initiated in 1992 after a National Workshop on purple loosestrife was held in Ottawa (Canada) to discuss the escalating invasion. The Manitoba Purple Loosestrife Project was formed shortly afterwards as a non-profit coalition between Agriculture Canada, the City of Winnipeg, the Canadian Wildlife Service, Environment Canada, Ducks Unlimited Canada, Manitoba Conservation, the Manitoba Naturalists Society, the Manitoba Weeds Supervisors Association, and the Delta Waterfowl Foundation. The list of stakeholder groups reflected the many environmental disciplines having an interest in the management of invasive plants. The partnership allowed for the sharing of resources, which was critically important since the project was never able to secure long-term dedicated funding. Initial efforts to control purple loosestrife spread began not in the field, but in back yards and gardens. A swap programme was developed in an effort to educate and encourage gardeners to destroy their purple loosestrife plants and 'Green Teams' were hired that provided free removal services to senior citizens and residents otherwise unable to dig out their purple loosestrife. The highly successful programme was supported by a wide range of media interactions (e.g. adverts on radio and television, articles in newspapers); the media became a partner and an important tool in the management response.

Given how widespread the species was, eradication and containment were not feasible and so it was necessary to focus on developing classical biological control within the context of an integrated management strategy (Lindgren 2000; Henne & Lindgren 2005). International partnerships were formed with Cornell University and the University of Minnesota (USA) to collect biocontrol agents from Europe, and to develop release and monitoring protocols. Mass rearing procedures were developed and summer students were hired to rear and release thousands of biocontrol agents across Manitoba. Partnerships were formed with regional rural municipalities, where these were provided with instructions and equipment to mass rear and release agents themselves, further empowering local plant management efforts. Innovative methods were employed, such as parachuting vials of beetles out of windows of fixed-wing airplanes to reach populations in remote wetlands.

One of the greatest challenges in controlling purple loosestrife was to address horticultural sales, as numerous cultivars had been developed for residential landscaping and gardens. These were being advertised as winter-hardy, ideal perennials for the home garden and excellent choices for perennial or mixed borders. The project played a pivotal research role in providing new data that demonstrated that cultivars previously considered sterile could easily cross with other plants, resulting in new plants that produced viable seed (Lindgren & Clay 1993). As a result of these new data, the project was able to convince the provincial government to revise the *Noxious Weeds Act* so that all varieties and cultivars of purple loosestrife were prohibited for sale. Previously, only *Lythrum salicaria* was regulated.

The project was able to employ a project manager (i.e. a champion) for almost 14 years despite most funding being through short-term grants (1–3 years). The partnership approach formed a model copied by other provinces across Canada.

8.2.2 Salt Cedar (*Tamarix* spp.)

The invasion by salt cedar is considered to be one of the worst ecological disasters impacting riparian ecosystems in the USA (DiTomaso 1998; DeLoach *et al.* 2000). In 1999, the Tamarisk Coalition was formed to address the invasion of riparian habitats by salt cedar in western Colorado (USA). The coalition has since expanded its scope to include all areas of the western USA, including over 40 partner organisations. Its strategic plan has three key objectives: first, to act as an information clearinghouse; second, to empower practitioners; and lastly, to enhance frameworks for restoration. The goals of the project are to enhance riparian stewardship, inform policy, adapt and integrate, foster partnerships, and foster knowledge. A milestone for the project was the publication of the 2006 *Salt Cedar Russian Olive Demonstration and Control Act* which has a focus on education and outreach. A strength of the coalition is its ability to work across jurisdictional boundaries, establishing cooperative weed management areas (see Box 8.1). It organises and hosts an annual conference, maintains a webpage, provides training, supports collaborative partnerships, delivers webinars, and monitors a classical biological control programme.

8.2.3 Leafy Spurge (*Euphorbia esula*)

Leafy spurge is an herbaceous plant introduced from Europe to North America in the eighteenth century. It has spread across pastureland and rangeland in the USA and Canada, decreasing forage production by as much as 75%, affecting and sometimes killing cattle (Lym & Kirby 1987). Leafy spurge has been a toxic thorn in the sides of farmers and producers in western Canada and the USA for decades. In response, TEAM (The Ecological Area-wide Management) Leafy Spurge ran from 1998 to 2004 with funding from the Agricultural Research Service and the Animal and Plant Health Inspection Service, US Department of Agriculture.

A primary goal of TEAM Leafy Spurge was to demonstrate the value of ecologically sound integrated pest management strategies by involving a wide range of partners (including various government agencies, universities, and private organisations) (Hodur, Leistritz, & Bangsund 2006). The programme consisted of five major components: programme management; operations; assessment; supporting research and demonstration; and technology transfer (Prosser *et al.* 2002). The project focused on managing leafy spurge by employing the principles of integrated pest management and classical biological control. To accomplish its task, TEAM Leafy Spurge stressed teamwork and assembled an experienced group of researchers and land managers into a focused, goal-oriented team.

8.3 Implementing Contingency Plans: Early Detection and Rapid Response

Contingency plans are designed so that when an event happens, preparation is in place, and a response can be rapid. Such plans should also include measures to improve detection. Watch lists (Fig. 6.2) are an important component of preparedness and should be supported by the availability of identification materials and alerts to support surveillance of species of concern. Pre-existing agreements on cost-sharing will facilitate implementation (see the Emergency Pest Plant Response Deed at www.agriculture.gov.au/biosecurity/partnerships/nbc/intergovernmental-agreement-on-biosecurity). The overall purpose of a contingency plan is to prevent the establishment of an invasive plant before it can begin to reproduce and spread, and as such increase the likelihood that a response will be effective (Wotton & Hewitt 2004). The implementation of contingency plans in the invasive plant management literature is often referred to by the term early detection and rapid response (EDRR), with EDRR often employed to refer to a whole range of different management approaches (Box 8.3).

> Box 8.3 *Invasive Species Early Detection and Rapid Response (EDRR): A Land Conservation Challenge for the Twenty-First Century (Randy Westbrooks & Steven Manning)*
>
> To date, world weed geographer Rod Randall from Western Australia has documented over 37 925 species of invasive plants around the world. Of this total, biogeographer John Kartesz from North Carolina has documented over 1300 introduced invasive plant species in the USA. Do you know how many introduced invasive plants already occur in your country or region?
>
> In response to this emerging global problem over the past 50 years, considerable effort has been made by state and national agencies, as well as other partners, to minimise establishment and spread of newly introduced and/or emerging invasive plants through three different approaches. This includes single agency-led programmes (EDRR 1.0), interagency councils and task forces (EDRR 2.0), and, most recently, the Landscape Approach to Early Detection and Rapid Response (EDRR 3.0).
>
> Two examples of successful single agency-led weed eradication programmes are the USDA Carolinas Witchweed Eradication Program in the USA, and the Kochia Eradication Project in Western Australia (EDRR 1.0). Over the past 57 years, through the efforts of a dedicated staff and seasonal employees, witchweed (*Striga asiatica* (L.) O. Kuntze), which is a root parasite of grass crops such as maize and wheat, has been reduced from a high of 432 000 acres in the eastern Carolinas to about 1200 acres at the present time. The total cost of the US Witchweed Eradication Program so far has been well over USD 250 million. However, it should be noted that this is a fraction of the annual economic

losses that would occur in national corn production if witchweed had been permitted to spread to major corn-producing areas of the Midwestern USA and Canada. If current funding of USD 800 000 per year is sustained, witchweed should be totally eradicated from the USA in another 10–15 years.

The introduction and subsequent eradication of kochia (*Bassia scoparia* (L.) A.J. Scott) in Western Australia is another good example of EDRR 1.0 in action. Kochia, which is native to Eastern Europe and Western Asia, was first imported into Western Australia in 1990 as a salt-tolerant pasture forage plant. Although palatable to livestock, kochia is poisonous in large quantities. It also alters fire regimes and reduces the diversity and abundance of native plant species. By 1992, the Western Australia Department of Agriculture recognised the weedy potential of kochia and targeted it for containment and eradication in 1993. Eradication strategies for kochia included burning, grazing, chemical control, and mechanical removal. By 1996, 46 of the 52 known populations had been eliminated from the state. The last sightings of kochia in Western Australia were in March 2000. In retrospect, it is obvious that the introduction of kochia was a short-sighted idea that cost at least AUD 500 000 to correct. However, this pales in comparison to future losses and control costs that would have been incurred by Australian landowners if the plant had been allowed to continue spreading.

In recent years, state and provincial interagency councils and task forces have been formed to address all types of new invasive species – particularly newly introduced species that are not already regulated by federal or state agencies. The Delaware Invasive Species Council, the Ontario Invasive Plant Council, and the Beach Vitex Task Force are good examples of this new trend in interagency partnering (EDRR 2.0). Beach vitex (*Vitex rotundifolia* (L.)f.), which is a woody vine that is native to the beaches of Korea, was first imported into the USA by the North Carolina State University Arboretum in 1985 as an ornamental for beachfront properties in the south-east. By the early 1990s it had started spreading to adjacent properties. By 1998 it was recognised as a threat to coastal dune restoration projects in the Carolinas. By the early 2000s it was also recognised as a serious threat to sea turtle nesting along the Carolina Coast. The Beach Vitex Task Force, which was first formed in 2003 to address this new invasive species problem, has achieved great success by eliminating most of the 230 documented beach vitex populations along the South Carolina coast. It is currently working with various agencies and homeowners to eliminate beach vitex from the beaches of North Carolina and Virginia.

From a societal standpoint, due to accelerating global climate change and increased global trade and travel, it is inevitable that the impacts of invasive species on food security, human health, and biodiversity will continue to increase unless steps are taken now to minimise their introduction, establishment, and spread. Development of EDRR capacity at the local level is the best way that landowners and managers can assist in meeting this land conservation challenge.

With this in mind, and considering current agency budget limitations in most countries, it is only prudent that local landowners and managers take steps now to protect their own natural and managed resources from all types of introduced invasive species, regardless of whether they are regulated or not. The Landscape Approach to EDRR – which involves development of EDRR capacity at all levels of the landscape, from local to national – is the newest EDRR strategy (EDRR 3.0). It includes individual public and private land units, geographic land units (e.g. watersheds, biomes, and corridors), and political land units (e.g. towns, counties, states/provinces, and nations).

Considering the ecological and economic impacts that we are already seeing from invasive species, it is imperative that we take steps now to protect future generations, native ecosystems, and agricultural production systems from this human induced biological tsunami. This can be done by strengthening our current invasive species management systems – i.e. Foreign Pest Prevention and Exclusion, Single Agency-led Pest Eradication Programs (EDRR 1.0), Interagency Councils and Invasive Species Task Forces (EDRR 2.0), the Landscape Approach to EDRR (EDRR 3.0) and by long-term efforts to control established invaders. It is very important that people take steps to prevent the establishment and spread of new invasive species on lands that they own and/or manage, for now, and future generations.

Partnerships Now! Weeds Won't Wait!

John Waugh (2009) published a book examining EDRR programmes against biological invasions along trade pathways in the USA, aptly called *Neighborhood Watch*. He suggested that the function of EDRR is to prevent the establishment of an invasive species before it can begin to reproduce. Waugh further suggested that there is a golden hour for EDRR, since once the invasion crosses a critical threshold, control becomes expensive and futile. In his report, he outlined key steps in the development of an incursion response programme: (1) have a well-developed capacity for detection; (2) have a formal reporting system; (3) have resources available to confirm species identification; (4) evaluate management options; and (5) evaluate the response (i.e. a treatment plan). Waugh argues that EDRR programmes should be associated with key control points (e.g. cargo containers, inland ports, airports) along trade-related pathways of introduction. He concludes by making a number of recommendations, including that EDRR can be enhanced at ports of entry by developing scenario exercises with port pest risk committees.

Preparedness for a response to an invasive plant incursion requires a number of key elements, including establishing a reporting structure, an agreed decision-making process, a response plan, and a coordinated response procedure (Biosecurity Queensland 2009). Planning for a response occurs at four levels – federal, provincial, regional, and local. Stakeholders include the federal and provincial governments, Weed Management Authorities, industry, and the community. A major issue with such planning is the need for resources to become available quickly, often in cases where there is no facility to roll over funds from year to year.

It is, of course, very well to develop a plan, but it is difficult to know in advance whether it will succeed. Specific exercises are needed to test the efficacy of potential control methods. There is also a need for emergency control exercises to test procedures and equipment, and training to ensure staff are well prepared.

8.4 Raising Awareness and Communication

Education and awareness have always played a significant role in supporting invasive plant management (Fig. 8.2). For example, the first federal weed publication in Canada was Lyster Dewey's 1893 pamphlet on Russian thistle, followed by James Fletcher's pamphlet, also on Russian thistle, in 1894, appearing in the *Winnipeg Free Press* (Evans 2002). In 1897, James Fletcher, the Dominion Botanist, published a bulletin on weeds (Canada Department of Agriculture Bulletin 28), and later in 1908 he, with George Clark, published *The Farm Weeds of Canada*. Over 100 pages long, *The Farm Weeds of Canada* was published under the direction of the Minister of Agriculture, Sydney Fisher, and included details on many common weeds at the farm gate, including ragweed, ox-eye daisy, ragwort, Canada thistle, hawkweeds, sow thistles, as well as illustrations of seeds.

Education and awareness activities are important in fostering social change. Targeted educational programmes that aim to educate gardeners

8.4 Raising Awareness and Communication · 211

Figure 8.2. The importance of education and outreach in supporting legislation was recognised very early. The first edition of *The Farm Weeds of Canada* was published in 1906, and included botanical drawings of various weeds.

on the impacts of invasive plants are essential. We have seen changes in society's attitudes towards smoking and wearing seatbelts, and a similar process is needed when it comes to thinking about invasive plants. Changing how society thinks about invasive plants will also lead to local political advocacy, which in turn translates into policy changes. There are a wide variety of such tools available (Box 8.4). Finally, it is important that the risk management options are communicated in a clear and transparent manner (Box 6.3). This includes exit strategies, as governments are often accused of walking away from problem pests when they get too costly. Management plans need to be clear and transparent as to decision switch points.

> Box 8.4 *Raising Awareness About Invasive Plants in Portugal (Elizabete Marchante & Hélia Marchante)*
>
> Citizens all over the world play a major role in introducing and spreading invasive plants, but they can also prevent, gather information on, and control such species. Therefore, when considering the management of invasive plants, the importance of raising awareness cannot be ignored.

In Portugal, even though biological invasions and their consequences have been recognised by law since 1999, a large proportion of the population is still unaware of them – some invasive species are still deeply admired by many for their ornamental characteristics, commercial value, or simply as features in new traditions, cultural habits, or gastronomy. By completely ignoring the problem, many people still facilitate the use of invasive species, aggravating the situation. This reality highlights the need for promoting public awareness and increasing engagement of the Portuguese public with invasive species. In this context, a team (including scientists from the Centre for Functional Ecology at the University of Coimbra and the Coimbra Agricultural School, Polytechnic of Coimbra) has devoted considerable effort over the last 15 years to raise awareness and to get citizens proactively involved in mitigation efforts.

Nowadays, many activities developed by our team are centred on a dynamic webpage that gathers information about plant invasions in Portugal (www.invasoras.pt) and includes a citizen science platform (online map & android app) where users all over the country can report sightings (Marchante *et al.* in press). In particular, since 2013 there has been a strong investment in the webpage, making it central to the efforts at raising awareness. The new webpage includes digital multimedia/interactive content and was set up not only to make background information available but also to provide updates, and get the public engaged in the mapping platform. This required professional support, but led to a fourfold increase in page views. Associated with the renovation on the webpage, in 2013 we started a public campaign in collaboration with a major national newspaper (circulation >100 000). This consisted of a series of articles published on Sundays for 12 weeks, leading up to an eight-page supplement on biological invasions. This had the advantage of reaching the general public, though it is unclear what its impact was.

To get more of the public participating on the mapping platform and using the website, while simultaneously increasing awareness on the subject, several outreach activities are ongoing. These include: (1) workshops (focused on the mapping platform); (2) summer fieldwork projects (including training, invasive plant control activities, and scientific experiments); (3) short training courses (Box 8.4 Fig. 1) for professionals dealing with alien plants (focused on identification and management and/or control, making use of webpage control videos); (4) short training courses for schoolteachers (focused on

8.4 Raising Awareness and Communication · 213

Figure 1. Technical short training course on identification and control of invasive plants in Portugal. Photo courtesy of Ana Pereira, CRE Porto.

awareness tools); and (5) projects with high-school students. While such hands-on activities have proven to be particularly successful in directly engaging people (Marchante & Marchante 2016, Schreck Reis *et al.* 2013), they also provide an opportunity for the distribution of printed materials about invasive plants in Portugal, including identification field guides, a technical volume about identification and control, bookmarks, and postcards (Box 8.4 Fig. 2).

More recently, social media (mostly www.facebook.com/InvasorasPt, but also https://twitter.com/Invasoraspt and www.youtube.com/user/InvasorasPT/videos) have been used to reach new audiences, working as a broadcasting hub and promoting interactivity between specialists and the general public. In contrast to practical activities, where people join mainly because they already have some interest in environmental problems, social media can be successful in reaching a wider audience.

Overall, public awareness about biological invasions in Portugal is increasing, and we believe our work has made a valuable contribution to it, but much more work is still needed. We assume that many people are now aware of invasive plants, have some affinity to environmental themes, are directly affected, or have been exposed through

Figure 2. A range of printed material about invasive plants in Portugal has been produced and distributed. Photo courtesy of Hélia Marchante.

school programmes and the like. However, a large majority of citizens is likely still unaware of the problem. Communicating about a negative theme (some of the beautiful plants and cute animals that we love can have negative effects after all) is quite a big challenge! The challenge to grow beyond the critical mass of peers and professionals and find allies in the common citizens is still daunting. Future work will involve diversifying the field actions, namely by establishing protocols with local and regional entities, promoting workshops for clubs at universities and public libraries and planning a pilot early-detection programme.

8.5 Recommendations

For an incursion response to be successful there needs to be an appropriate organisational structure set up, strong leadership, and ongoing partnerships, e.g. through working groups and oversight committees. These should be created in advance of the problems and not simply in response to them. Awareness raising and communication need to be closely aligned to on-ground management, since implementation will only be successful in the context of a broad societal appreciation of the importance of managing biological invasions.

Processes	Requirements	Deliverables
Develop and maintain appropriate organisational capacity	• Legal mandate • Estimation of geographical and temporal scope of response needed • Availability of funding linked to specific roles and activities • Inspectors in place with mandate to enforce laws and regulations • Information management systems (including data collection standards and forms)	• Champion • Team capable of acting • Reports on programme progress
Develop partnerships	• Understand differences in mandates of organisations • Appreciation of skills required • Resources for coordination, ideally dedicated sustainable funding • Frameworks for communication and data sharing • Well-defined expectations and responsibilities • Dispute-resolution mechanisms	• Memoranda of understanding, and funding agreements • Agreement with specialists for consultation purposes • Regular workshops and planning meetings • Increased coordination between jurisdictions • Appropriate scientific oversight of progress
Ensuring preparedness for incursions	• Horizon scanning • Taxonomic and identification capabilities • Prioritisation of future risks • Mechanisms to ensure control operations can be funded and implemented rapidly	• Watch lists • Contingency plans • Emergency response capacity

(cont.)

Processes	Requirements	Deliverables
Improving public awareness and support	• Communication strategy • Dedicated funding and coordination	• Dedicated website and social media tools • Publicity material • Educational materials and tools

9 · Conclusions and Future Directions

9.1 Overview

Below we consider key conclusions from each of the chapters, identify critical issues and suggest major areas for future research and development (Fig. 9.1).

9.2 Predictions

Efforts at prediction have largely focused on species, as opposed to areas or pathways, to the extent that the search for a weedy phenotype has been somewhat of a Holy Grail of invasion science (Sections 2.1–2.3). While it is possible to form generalisations that are highly useful in constructing hypotheses and furthering our understanding, they provide little basis for prediction per se. A few common correlates of invasiveness and impact (e.g. native range size) seem to be fairly robust, and might be indirectly linked to invasive success and impact. For other traits there are clear mechanistic links to invasiveness and impact (e.g. the greater the propagule pressure the more chances for invasion, and invasive plants that grow taller than native vegetation tend to have large impacts). Correlates and traits provide useful first approximations, but that is all. To date, most predictions of value to management are specific to particular invasion stages, environments, phylogenetic groups, or management contexts – except that if a plant is invasive and has impact somewhere it is likely to become invasive and have impact in other areas that have similar climates.

Another complication is that the traits that result in impact can be different from those that determine the success of an invasion (Section 2.3). Impact is difficult to predict and will take longer to be observed (too long to be of use to inform proactive management). The quantification of existing impacts is a complex undertaking, since there are many aspects, both negative and positive, to be considered (Table 2.1, Box 2.1). Efforts

218 · Conclusions and Future Directions

(a)

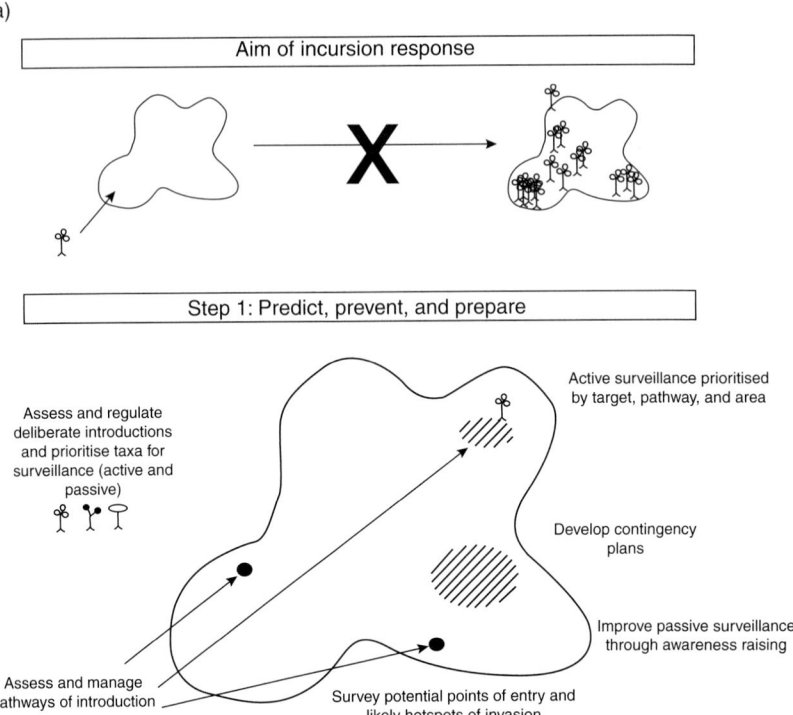

Figure 9.1. A summary of incursion response. The overall goal of incursion response is to limit the number of widespread invasive taxa that require ongoing maintenance management. We consider three main steps that need to be taken. These will also require several facilitating mechanisms and activities.

at defining impact are ongoing, and are looking promising. For example Ricciardi et al. (2013) have identified a number of testable hypotheses that explain temporal and spatial variation in impact. An ability to quantify invader effects, at least roughly, would enable managers and decision makers to make a regulatory decision and improve prioritisation, and the IUCN Environmental Impact Classification for Alien Taxa Scheme should help (Blackburn et al. 2014; Hawkins et al. 2015). Providing accurate predictions of economic impact is particularly problematic at present (Box 7.2), but order-of-magnitude estimates should be sufficient for the purposes of deciding whether or not to mount and sustain an incursion response.

Finally, to manage invasion debt effectively, it is vital that we can predict how distributions will change over time (Section 2.4.2). Many techniques

(b)

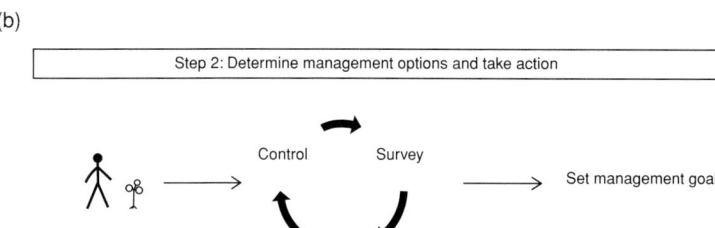

Process	Information requirements	Deliverable
Initial incursion response	• provisional identification (with formal identification to follow) • evidence of non-native origin • basic details of known naturalised distribution (including field notes) • basic risk assessment	• physical specimen in a recognised collection • naturalisation report (if appropriate) • documented management decision with conditions under which the decision should be revisited
Determining feasibilities of eradication and containment	• basic biology (e.g. reproductive ecology...) • value and interactions with humans • list of potential targets for management • current distributions • historical route and potential future pathways of introductions • estimates of the probabilities of new introductions and of there being other as yet undetected incursions • seed or propagule bank size • habitat suitability or risk map • dispersal pathways at various scales • detectability • management efficacy • control and project costs • photographs of different stages and control operations • bioeconomic or decision support model • cost–benefit analysis	• priority list for management
Strategic framework for incursion response	• estimates of likelihood of achieving management goals under different levels of funding	• strategies favoured under different resourcing • decision switch points

Figure 9.1. (cont.)

exist for species distribution modelling, and there is a substantial literature dealing with methodologies (Box 2.3). At a broad scale, species distribution models can be employed to estimate the total area potentially at risk of invasion which, together with estimates of dispersal pathways and potential impacts in different land-use types, can provide a quick, broad-brush map of the risk posed by a particular incursion (Fig. 2.3 Box 3.1). With continued advances in geospatial information systems and computer algorithms, species distribution modelling should continue to be a fruitful area for research, particularly in the context of a changing climate. A major challenge will be to incorporate dispersal into such models. At present, spread rate is commonly considered via an inspection of possible

220 · Conclusions and Future Directions

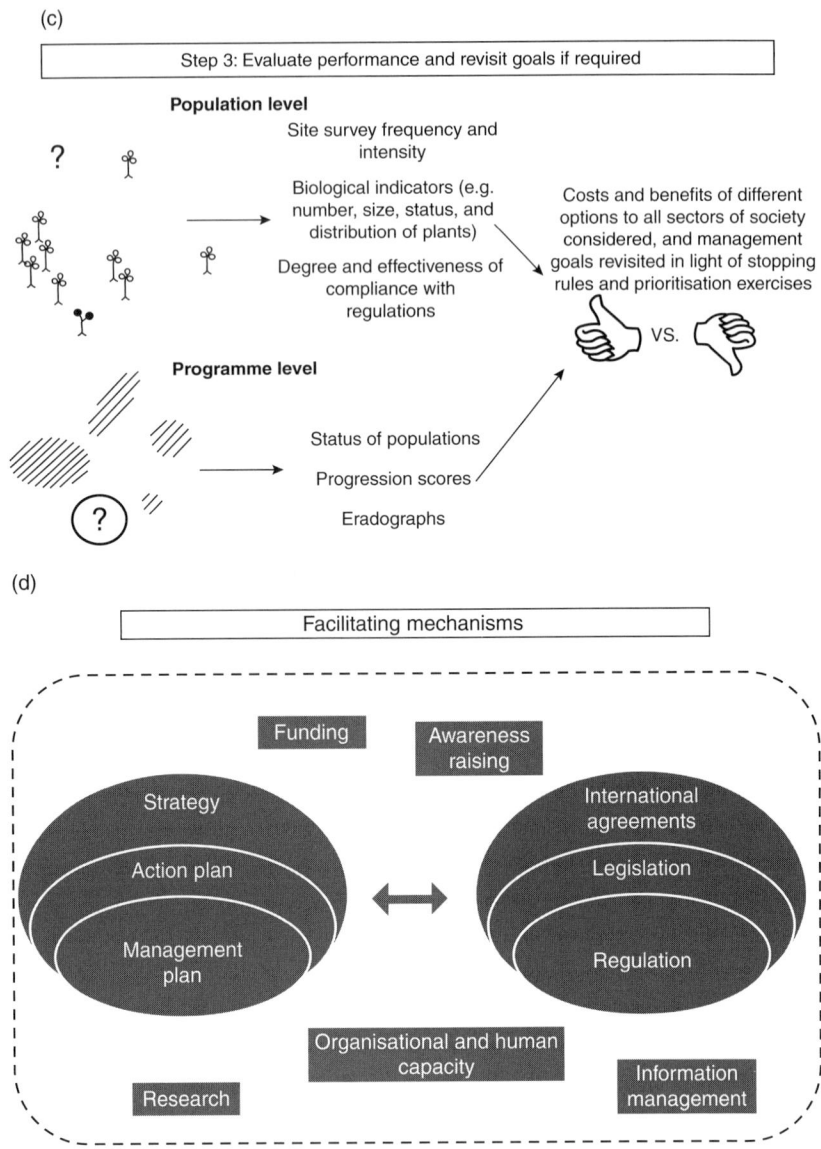

Figure 9.1. (cont.)

spread scenarios, with scenario selection largely a function of perceived hazards. An improved mechanistic understanding of the invasion process should help identify those invaders that are likely to spread fastest and cause the greatest damage.

9.3 Detection and Delimitation

If border prevention has failed then detection post-border becomes critical. A better understanding of pathways of invasion, comprising both their multiplicity and relative importance, will help (Section 3.1, Fig. 3.1). Current patterns of invasion are the result of very specific processes set in place decades or centuries ago (Section 3.1). However, novel pathways, such as e-commerce and the use of invasive plants for biofuel (Box 6.2), are a challenge. As such, the prediction of new incursions can be improved by considering historical factors and will often need to take social history into account. Similarly, future planning needs to explicitly consider current processes. More work is required on methods to quantify the relative importance of multiple invasion pathways, but also on how pathways are amenable to management intervention. Some pathways might generate effective long-distance dispersal, but they can also be regulated, and often managed (Table 2.2). Other pathways, however, will have to be dealt with either at source or at their likely destinations. In any event, it is important to have effective surveillance at potential invasion hotspots (Section 3.2).

At present, if an incursion happens to be detected at a stage where eradication is feasible, this is generally a result more of good luck than of effective procedures. Therefore a priority for most countries is to improve surveillance capacity and networks, in terms of employing dedicated capacity to survey and identify taxa, but also by facilitating input from the wider public, e.g. by running citizen science initiatives, awareness campaigns, and interacting with stakeholders (Section 3.3). Detection efforts are significantly enhanced by employing citizen scientists who are trained in detection and identification techniques. The more eyes in the field, the better the chances of detection.

When an incursion is detected, delimitation is essential if management options are to be properly evaluated (Section 3.6). If the potential area invaded is small and resources permit, the whole area should be surveyed in detail. However, very large areas are often at risk of invasion and so search activities must be prioritised. Species distribution models can be used to target surveillance efforts, saving time and resources (Box 3.1). More research is needed to determine how easy it is to find plants in a given area (Section 3.5, Box 3.2). Improved estimates of detectability are required for different growth forms in different types of vegetation, so that reasonably accurate, first-pass estimates of required search effort can be generated. While it is possible to get a qualitative feel for detectability during control operations (i.e. field experience), few studies have quantified detectability. Focused research on detectability could provide a quantum improvement in incursion responses.

Technological innovations have the potential to radically improve detection and delimitation activities over the next few years. Smartphones with applications such as electronic field-guides and easy reporting functions have improved the detection rates achieved through weed spotter networks. Molecular identification tools have provided valuable additions to traditional morphological identification, with DNA barcoding already being used by some biosecurity organisations. Remote sensing is being used by many organisations to detect the spectral signatures of invasive plants, and Google Earth can provide valuable baseline information (Visser *et al.* 2014). Looking towards the future, drones will likely be used increasingly to scout for plants. However, the potential of such techniques still needs to be realised, and 'boots on the ground' and 'eyes in the field' remain vital. While passive surveillance is a very important component of detection and delimitation, it will be essential to retain qualified professionals who spend time out in the field as part of the overall surveillance strategy (Section 3.5, Box 3.2).

9.4 Evaluating Management Options

Once an incursion has been detected, it is important to decide on an appropriate response (Section 4.1). Often a decision will be taken not to manage an incursion. All decisions must be properly documented, including conditions under which they would be revisited. Assessing management options can be done using a combination of quantitative, semi-quantitative and qualitative models, as dictated by the available data (Section 4.3, Tables 4.3 and 4.4). However, the political environment, public pressure, and the spatial extent of an incursion all play significant roles in determining whether or not eradication is a feasible goal (Section 4.4).

During the past 15–20 years there has been substantial field-based research into eradication feasibility (Section 4.3.1, Fig. 4.4, Box 4.1). The results obtained have supported earlier views on the limitations to eradicating plant incursions. However, such views have been described as 'pessimistic', and there are systematic issues with such analyses. Perhaps the most important lesson has been the need to allocate sufficient resources to delimitation, since the gross area appears to set the upper boundaries for what is possible, subject to the particular biology and ecology of the targeted species (Section 3.6).

In all cases decision makers should be provided with estimates (however rough) of the probability of achieving any management goal, along with functions describing how such probabilities might vary according to the level of investment made. It is then up to them to select the probability of success that is acceptable, considering the available resources (Section 4.4). Lower probabilities might be entertained where higher levels of risk are concerned. It should be noted, however, that there is a significant risk associated with the commitment to a coordinated incursion response. This is related to the so-called opportunity cost of investment, in that resources invested in one programme are by definition not available to others. Some surprising results might arise from the formal consideration of the threshold probability of success (Fig. 4.8). For example, through a cost–benefit analysis relating to the eradication of an invasive tunicate species, Edwards and Leung (2009) demonstrated that an eradication attempt would be economically worthwhile as long as there was more than a 16% chance of success. Clearly, more work needs to be undertaken in this area to better understand the economics of incursion responses.

9.5 Evaluating Management Performance

Management performance should be evaluated at both a population level (Sections 5.1 and Sections 5.3.1), and a programme level (Sections 5.2 and 5.3.2). Performance towards achieving eradication is measured in terms of abundance and persistence at one or more sites, and performance towards achieving containment in terms of area invaded. There is also significant value in considering performance against goals in combination (e.g. eradographs, Section 5.4, Fig. 5.9). These procedures for evaluating management performance are powerful tools, but can be applied only if operational staff collect the essential, albeit basic, data. Protocols for collecting data are readily available but will inevitably need to be adapted according to experiences in the field (Figure 5.1).

Importantly, an eradication programme that does not achieve its primary goal will not necessarily represent significant wasted investment, as it will likely have reduced spread. However, there will be a point at which the primary goal of an incursion response must change (Section 5.5). Decisions for any changes to the course of management must be informed by what is happening on the ground in the context of modelled expectations. Decision makers need to make a call on whether to increase investment, opt for another management goal, or abandon coordinated management in favour of maintenance management. Similarly, applied ecologists need to keep assessing whether the decision support models produced are appropriate, or if they are missing some key aspects of the true dynamics of the system. A regular cycle of evaluation and review is required. This needs to consider metrics appropriate both at the scale of individual populations and at a programme level (Sections 5.1–5.4). Since eradication has a more clearly defined goal, decision making is relatively easy compared to determining whether and when to cease containment efforts. More research is required to define stopping rules for containment, and determine how to translate decision switching points (currently defined in bioeconomic terms such as 'net benefit') into forms that can be calculated and applied in practice.

9.6 Legislation and Agreements

Legislation plays a critical role in facilitating the prevention, detection, and management of plant incursions (Boxes 6.1 and 6.2). It can take various forms at various spatial scales (Section 6.2). International agreements entered into by National Plant Protection Organisations and biosecurity agencies also play significant roles in protecting biological diversity, plant health, and trade (Sections 6.1 and 6.3). The requirement for science-based analysis of risk is increasingly embedded in legislation. One of the chief tools used by regulators is to list species in different categories that demand particular actions (Section 6.4). Ensuring this process is transparent and that the categories chosen are the most appropriate will be an important area for future policy work.

Traditionally, legislation concerning alien plants has been reactive, but we are witnessing a paradigm shift. Many authorities have begun to regulate pre-border and industries are beginning to self-regulate (Section 6.5). There are, of course, many evolving challenges (Section 6.6), and there needs to be a more careful assessment of the advantages and disadvantages of different approaches, from a heavy hand to a light touch (Section 6.6.5). Finally, we note that the likelihood of successful coordinated management depends both on support from (and for) stakeholders (Boxes 6.2 and 6.3). Funding for this is often needed over relatively long time scales.

9.7 Strategies and Actions

Strategies against invasive plants are developed and implemented at various spatial and temporal scales and by a wide variety of organisations and people (Box 7.1). There are many approaches to strategy development and plenty of existing examples to study (Section 7.1, Table 7.1, Figs. 7.1 and 7.2). These considerations point to some general rules. A broad, diverse set of stakeholders should be engaged early in the process to ensure buy-in (Section 7.2). All stakeholders that are directly and indirectly impacted by invasive plants should be consulted, taking into account relevant international standards and stakeholder concerns (Fig. 7.1). Crucially, in such engagements there needs to be a protocol for breaking stalemates; disagreements should not lead to inaction by default. A strong champion is required (Section 7.3), and the champion must be supported with the appropriate level of resources to get the job done, whether this is individually or, ideally, as part of a group comprising representatives from the leading partners. Champions are the reason many strategies are successful, and the lack of a champion is often why others fail. The goals set must be specific, measurable, attainable, realistic, and time-bound (SMART) (Section 7.4). The roles and responsibilities of the partners involved must be clearly identified and agreed upon. This can be one of the most challenging aspects in strategy development, as partners and stakeholders might be reluctant to sign on to a strategy if there are resource implications.

A high-level strategy needs to consider how to allocate resources to the variety of tools and approaches available (e.g. van Wilgen *et al.* 2011). We would caution, however, that there needs to be a mechanism to ensure resources are ring-fenced for incursion response, since this is often squeezed between the competing demands of pre-border prevention and high-profile campaigns against widespread weeds.

Finally, a strategy should not just be something that can be printed, bound, and placed on a shelf. Strategies are often technically feasible but lack sufficient funding or coordination for implementation. All agencies, and not just the lead agency, must be adequately supported. National strategies are rarely revisited or updated, so mechanisms are needed that allow for re-evaluation and adaptation. It is imperative that resources are available both to implement a strategy and to assess its performance. Often it will be necessary to have separate action plans where the details of how a strategy will be implemented are outlined.

9.8 Implementation

Collaborative partnerships are required to detect and manage plant incursions and this will almost invariably involve multiple diverse stakeholders

(Sections 8.1 and 8.2, Boxes 8.1–8.4). Collaborative partnerships that pool resources and funding are almost always necessary to mount a successful incursion response.

There needs to be significant thought as to what to do before an incursion happens (i.e. contingency planning, Section 8.3), and when it does happen there needs to be emergency capacity to identify the problem and deal with it (e.g. an early detection and rapid response (EDRR) programme, Box 8.3). There must also be a champion to ensure that, when underway, programme funding can keep flowing and project targets are met. Finding ways of maintaining the enthusiasm and diligence of workers and volunteers will tax even the savviest of team managers, particularly as targeted plants become increasingly scarce as eradication is approached.

Prior to the implementation of an action plan, public information programmes that foster awareness need to be considered (Section 8.4, Box 8.4). Public cooperation is paramount when detecting and responding to incursions, and conversely resistance to an incursion response can render management goals unachievable.

9.9 Closing Remarks

We believe that significant improvements in the management of plant incursions would be made if there were better links between theory and practice (Fig. 9.2). An adaptive approach to the challenge will increase the applicability and uptake of theoretical models, as well as lead to the development of models that are more practical. Research should not delay management, and management should not be haphazard; research and management should be a single process.

Rather than letting the lessons of the past define the future, we should learn from them and, perhaps, aim higher (Simberloff 2002; Mack & Foster 2009). A downside of aiming higher is the potential for funder burn-out (Parkes & Panetta 2009). While failed eradication attempts might be worthwhile on balance, they are not particularly good for publicity. This means that more effort should be devoted to detection and surveillance, establishing the feasibility of eradication and containment, careful monitoring and evaluation of performance, and perhaps most importantly, calibrating expectations to the time frames necessary to achieve coordinated control.

If incursion response is successful there should be fewer invasions. We would like to be able to take our children and grandchildren to the grasslands in Queensland and show them that there is no pom-pom weed. To

Figure 9.2. Theory and practice do not meet often enough. Cartoon courtesy of Sindiso Nyoni.

the prairies of Canada that contain no saltcedar trees. And to the Kruger National Park in South Africa free from kudzu vine. If we want future generations to be able to decide whether they want native or alien plants in their region, we will need to manage incursions proactively. It is not actually that difficult, but it does require coordination and persistence. We hope this book can, in some small way, contribute to this vision.

Glossary

Some of the terms in this book are used in much wider contexts than described here, but we have tried to be precise. For terms that have substantially different alternative definitions (in particular: United Nations Environment Programme 2010; IPPC-FAO 2013a), we have made a note in italics. Please also note the definitions used by text box authors are at their discretion, so there might be some differences. See p. xviii for a list of abbreviations.

Action plan: a series of defined actions that will be taken to achieve specified goals, often outlined in a **strategy**. The **action plan** should include timelines, responsibilities, and budgets. Most **strategies** will call for action plans, and then the action plans in turn call for **management plans** which are directed at individual species, areas, or pathways.

Active surveillance: see **surveillance**.

Alien plant: a plant taxon in a given area whose presence there is due to intentional or accidental introduction as a result of human activity (synonyms: exotic plants; non-native plants; non-indigenous plants) (Richardson *et al.* 2000b). Species that have part of their native range in a given country, but whose presence in another part of the same country is attributable to human actions that enabled the species to overcome fundamental biogeographical barriers, are sometimes termed extra-limitals. The need for this additional definition arises because biogeographical barriers for species often do not match political boundaries (i.e. a species can be both alien and native in the same country). From an ecological standpoint, such species are similar to other alien or invasive alien species, but the notion of native species also being 'alien' gives rise to political and practical challenges.

Area-based invasion debt: see **invasion debt**.

Asset protection: a management goal whereby particular assets are identified and measures are put in place for their protection.

Glossary · 229

Barrier zone (also called buffer zone): part of the **containment area**, where measures are put in place to detect and control a targeted species (cf. **containment core**). Its purpose is to prevent further spread by preventing or minimising the amount of dispersal beyond its outer boundary.

Biological invasions: there is an ongoing debate about how to define biological invasions. This debate is largely moot, as there are three distinct concepts that are important – biological invasions as a biogeographical phenomenon, in terms of ecological dominance, and as interpreted through human values. For often seemingly good reasons these definitions are frequently conflated, and this has led to confusion at the interface between management, science, and policy. Throughout this book, we use biological invasions in the context of the biogeographical definition as laid out by Blackburn *et al.* (2011).

Biosecurity: measures that are taken to stop the introduction or spread of organisms harmful to human, animal, and plant life. In the context of this book such measures consist of a combination of processes and systems that have been put in place by governments and land managers to prevent the introduction, establishment, and spread of potentially harmful invasive plants.

Blackburn scheme: a scheme for categorising alien species based on how far it is along the introduction–naturalisation–invasion continuum in a specified spatio-temporal context. It consists of 11 alpha-numeric categories (A1–E) (Blackburn *et al.* 2011). For an example of how to apply it in practice to introduced trees, see Wilson *et al.* (2014).

Condensed area: see **net area**.

Conflict species: species that have positive value to certain people or groups, but negative value to others. In the context of biological invasions, these are often species that were deliberately introduced for a particular use (e.g. ornamental or production), but that subsequently became invasive and caused impacts. There is often a spatial and temporal separation between benefits and impacts, leading to disagreements about how these species should be managed.

Containment: the aim of preventing or reducing the spread of invasive species, e.g. by preventing incursions into new areas, and extirpating any species that are found outside a defined area or beyond a defined line. One of the four main **management goals** of invasive species management. Full (or complete) containment is when all spread beyond a specified **containment area** is prevented. Partial containment is when spread beyond a specified **containment area**

is significantly reduced by comparison to the spread that would occur in the absence of containment measures.

Containment area: an area that encompasses one or a number of populations, plus surrounding barrier or buffer zones, that are subjected to periodic search and control efforts.

Containment core: part of the **containment area** (cf. barrier zone) where control efforts might or might not be implemented. If they are implemented, this will generally be to a lower degree than the control that is exerted in the **barrier zone**.

Containment feasibility: the degree of difficulty anticipated in an attempt to contain an incursion. It is influenced not only by biological factors related to the species, but also the socio-political, economic, and operational factors specific to the context of the incursion (see also **eradication feasibility**).

Coordinated control: management programmes with the goals of either **eradication** in or **containment** to a specified area.

Delimitation: the process of determining the full extent of an incursion.

Detection threshold: the size or stage at which plants in an area are consistently found from a specified distance. The detection threshold can often be decreased by more intensive or more frequent surveys, but in some cases it is very difficult to detect and identify plants when they are small or immature.

Dominance: the last stage of the invasion process, where an invasion begins to reach high local abundance and starts to develop relatively stable range margins in its new range. By this point, the overall goal is often **asset protection**, though classical biological control can in some instances substantially reduce the abundance and extent of the invasion. Most management effort is spent on this stage, i.e. is reactive (cf. **pre-introduction**; **incursion**; and **expansion**).

Early detection and rapid response: a management approach where-by resources are spent to improve surveillance, and emergency measures can be implemented quickly in order to start dealing with an incursion once it is detected. *Note that EDRR is frequently used as a term to describe the range of management actions taken against species that are in the **incursion** or **expansion** stages, and is often described as a management goal in itself. However, in many cases control at these stages of invasion will require continued surveillance and sustained responses, and using the term EDRR in this context can create misleading expectations.*

Ecosystem transformers: a subset of invasive plants that change the character, condition, form, or nature of a natural ecosystem over a substantial area (Richardson *et al.* 2000b). These plants, comprising a small percentage of all invasive plants, have profound effects on biodiversity and ecosystem function, and should be **prioritised** for coordinated control.

Eradication: the elimination of every single individual (including propagules) of a species from a defined area in which recolonisation is unlikely to occur (Myers, Savoie, & van Randen 1998). If there is substantial uncertainty about the potential for recolonisation, the term **extirpation** should be preferred. If eradication is achieved there will be a significant change in the management action (cf. **extirpation**). *Note that specification of the likelihood of recolonisation is not used in some other definitions (e.g. the IPPC definition).*

Eradication feasibility: the degree of difficulty anticipated in an attempt to eradicate an incursion. It is influenced not only by biological factors related to the species, but also the socio-political, economic, and operational factors specific to the context of the incursion. Incursions with lower eradication feasibilities will require more resources if eradication is to be achieved. The term is sometimes used in a narrow sense (i.e. just considering the biological characteristics and environmental context), which can be referred to as the technical feasibility, but we generally use it in the broader sense that incorporates all aspects that affect the likelihood of achieving eradication (including the socio-economic context).

Eradication impedance: the factors slowing or halting an eradication attempt. The four groups of impedance constraints addressed in Panetta and Timmins (2004) were accessibility, detectability, weed biological characteristics, and control effectiveness.

Expansion: the third stage of invasion, where a plant invasion is increasing its range, sometimes rapidly. This is a stage at which the value of **coordinated-control** becomes questionable, but often **containment** can have significant value in reducing the rate at which the problem escalates (cf. **pre-introduction**; **incursion**; and **dominance**).

Extirpation: the elimination of every single individual and propagule of a species from an area in which the possibility of recolonisation cannot be ignored in practice (cf. **eradication**).

Gross area: the area that must be searched in order to delimit an invasion (cf. **net area**). In the main this area can be estimated only roughly, since it requires a knowledge of where dispersal is likely to occur to, as well as the habitats in which establishment is likely (i.e. habitat suitability). As with the **net area**, it is important to state how the area was calculated.

Impact-based invasion debt: see **invasion debt**.

Impact reduction: the goal of reducing the negative impacts of introduced species. This is often attempted while trying to maintain any benefits of an introduced species to an area (see **conflict species**). In some cases a management threshold is set, below which no further control is required.

Introduction debt: see **invasion debt**.

Incursion: one of the four main stages of the invasion process (cf. **pre-introduction**; **expansion**; and **dominance**; Fig. 1.2). Owing to the limited spatial extents generally involved in incursions, this is a stage at which **coordinated control** (particularly **eradication**) might be feasible. As such, species at the introduction, naturalisation, or early spread stages might be considered as incursions, i.e. B1–D2 but not E under the **Blackburn scheme**. *Note: according to the IPPC an incursion is an isolated population of a pest recently detected in an area, not known to be established, but expected to survive for the immediate future.*

Incursion response: the process of decision making and, when appropriate, controlling invasive and potentially invasive alien plants after they have been introduced but before they become so widespread that other management goals (e.g. impact reduction) are more appropriate. The most appropriate incursion response is often not to act further. Any such decision must be explicit and documented, ideally with conditions identified under which the decision would be revisited.

Invasion debt: the potential increase in the biological invasion problem that a given region will face over a particular time frame in the absence of any strategic interventions (Rouget et al. 2016). It is composed of the number of new species that will be introduced (**introduction debt**), the number of species that will become invasive (**species-based invasion debt**); the increase in area affected by invasions (**area-based invasion debt**); and the increase in the negative impacts caused by introduced species (**impact-based invasion debt**) over some specified time horizon and assuming current processes continue. Only a small proportion of the **introduction debt** will contribute to the **species-based invasion debt**, and likewise a small portion of the **species-based** and **area-based** invasion debts will contribute most of the **impact-based invasion debt**.

Invasion science: the study of all aspects of biological invasions and their consequences, including biology, ecology, economics, ethics, sociology, and inter- and transdisciplinary studies (Richardson 2011).

Invasive plants: naturalised plants that produce reproductive offspring, often in very large numbers, at considerable distances from parent plants (approximate scales: >100 m; <50 years for taxa spreading by seeds and other propagules; >6 m/3 years for taxa spreading by roots, rhizomes, stolons, or creeping stems), and thus have the potential to spread over a considerable area (Richardson et al. 2000b). *Note: in terms of the CBD (www.cbd.int/idb/2009/about/what/) there is an explicit recognition that such invasions can have undesirable consequences: 'Invasive alien species are plants, animals, pathogens and other organisms that are non-native to an ecosystem, and which may cause economic or environmental harm or adversely affect human health.' As such, the term invasive has been used in the context of native species and has sometimes been reserved for taxa that have demonstrable impacts, but in the context of this book only the biogeographical definition is used.*

Lag phase: a phenomenon whereby population growth or spread is initially slow, but increases rapidly later due to some change in population dynamics. The term **lag phase** is often used in contexts where there is a large rise in absolute abundance following a long period of low abundance; however, when such a pattern results from a small initial population size and a constant population growth rate the population was never technically in a lag phase. A related term is that of *sleeper weeds*. These are weed populations where there is an identified mechanism preventing rapid population growth or spread (such a population is often referred to as being in a **lag phase**). This is problematic, since often the mechanisms are not identifiable until after the weed 'wakens' and as such the term is of little practical value. It also suggests that 'sleepiness' is an attribute of the plant, whereas a lag phase is often attributable to the environmental context.

Management goal (syn. management objective): the endpoint that management is aiming to achieve. In the case of the management of biological invasions various schemes consider different sets of goals. In this book we consider four main high-level goals – prevention, eradication, containment, and impact reduction. *Note: schemes sometimes differentiate objectives to deal with the later stages of an invasion, with the inclusion of asset protection (e.g. Auld & Johnson 2014). Moreover, as conservation is often the overall rationale for the management of invasive plants, it is often important to explicitly consider restoration as a goal either during or following control.*

Management plan: a document that outlines the tasks, responsibilities, and budgets needed to achieve a stated **management goal**. The plan can refer to a particular area, pathway, or species and is often quite specific and prescriptive. It is often the document used to implement a specific part of an **action plan**, which in turn is used to implement the overall **strategy**.

Naturalised plants: alien plants that sustain self-replacing populations for many life cycles without direct intervention by people, or despite human intervention. Naturalised species are not necessarily invasive, that is they have not (yet) spread any significant distance. As such, dispersal is the key factor that allows naturalised plants to become invasive.

Net area: (syn. condensed area) (cf. gross area): the area actually covered by plants, and hence that needs to be treated. There are various methods for measuring condensed area, e.g. by calculating the total canopy area covered by the invasive plants, or by using alpha-hulls on point data, and so the measure used for calculation should always be stated.

Passive surveillance: see **surveillance**.

Pathway: the combined processes by which species are introduced from one geographical location to another (cf. **vector**).

Permitted list: a list of species that are allowed to be imported. In some cases it refers to species that are allowed in without conditions, and in other cases the species require permits.

Phenology: the observed state of an organism, the result of genetic and environmental influences. A plant's reproductive phenology often varies cyclically and in response to seasonal natural phenomena.

Phytosanitary measures: the steps taken to ensure that unwanted organisms (that might affect plants) are not introduced to an area.

Plant health (syn. plant protection; plant quarantine): term used to refer to the legislative and administrative procedures used by governments to prevent pests from entering and spreading within their territories. In North America the terms **plant health** and plant protection are used, while in Europe plant quarantine and **plant health** cover the same approach (Ebbels 2003).

Plant invasion: see **biological invasions/invasive plants**.

Pre-introduction: a stage in the invasion process, where a species is not currently present in the region of interest (it might have been introduced previously and either failed to establish or persist, or was eradicated). In this case prevention is the main management goal (cf. **initial incursion**; **expansion**; and **dominance**).

Prioritisation: the process of focusing management efforts, whether for detection or control, on a subset of areas, pathways, or species where it is considered that efforts will be most effective.

Prohibited list: a list of species that are not allowed to be brought into a country, or when used post-border, a list of species for which certain activities are not allowed (e.g. sale, spread, or cultivation).

Propagule pressure: a composite measure of the number of individuals introduced to a region, incorporating the absolute number of individuals involved in any one release event and the number of discrete release events (Lockwood, Cassey, & Blackburn 2005).

Quarantine: the processes and/or physical structures put in place to prevent the spread of an organism. **Quarantine** is generally required to achieve the management goal of **containment**. The word 'quarantine' originates from the Venetian dialect form of the Italian *quarantina*, meaning 'forty days'. To prevent the spread of diseases related to the Black Death, ships and people prior to entering the city of Dubrovnik were kept in isolation for a *quarantina* (i.e. 40 days).

Regulation: a law, rule, or other order prescribed by authority, especially to regulate conduct.

Risk analysis: the process of identifying, assessing the likelihood and consequence of an event, as well as developing strategies to manage and communicate the risks.

Risk assessment: the process of analysing and evaluating the likelihood and consequence of an event taking place. In terms of biological invasions this can be for individual species, pathways, or areas. **Risk assessment** is part of **risk analysis** (Chapter 4). **Risk analysis** comprises **risk assessment**, **risk management**, and risk communication.

Risk management: the process of assessing options by which the risks of an event (either its likelihood or consequence) can be reduced or mitigated.

Species-based invasion debt: see **invasion debt**.

Species distribution model (syn. ecological niche model): a model used to describe the current and potential environmental distributions of species that when projected onto real landscapes provide an indication of where species are expected to occur. Such models are often based solely on climatic variables, but can include estimates of other aspects of habitat suitability (e.g. soil variables and the presence of standing water).

Spotter network: a network of volunteers who report, collect, identify, and deliver specimens of potential, new, and emerging weeds in their region. Spotters can be landholders, gardeners, or members of community groups, along with government officers, industry representatives, and anyone else interested in weeds and plants. Such a network might involve coordinators, who help identify and filter weed spotters' specimens and pass them on to herbaria for formal identification and curation (cf. **surveillance**).

SPS Agreement: the Agreement on the Application of Sanitary and Phytosanitary Measures, an agreement as part of the World Trade Organisation whereby trade can be restricted in order to protect human, animal or plant life or health (www.wto.org/english/tratop_e/sps_e/spsagr_e.htm)

Strategy: a high-level plan for achieving management goals in a specific time frame under conditions of uncertainty. *Note: There is a wide range of definitions for a strategy. For a discussion of other definitions, see Chapter 6. A strategy will often also include an **action plan**, but it can often be useful to separate the two explicitly.*

Surveillance: a process that involves collection and recording of data on plant presence or absence by survey, monitoring, or other procedures. **Active surveillance** is where there is effort specifically dedicated to look for a particular plant or (for general surveillance) to look for all invasions in a given area. It generally involves planned, systematic searches of locations where targeted species are considered likely to occur. **Passive surveillance** is a form of surveillance whereby detections occur largely by chance. Resources and information may be provided so that if people encounter plants, the information can be documented and passed on to someone to act on it (e.g. through a **spotter network**).

Vector: a mechanism responsible for the transport of species to new areas where they did not previously occur. A **pathway** (for example shipping) could have several vectors associated with it (for example in cargo, in passenger luggage, on passengers or crew themselves, in ballast water, or attached to the hull).

Watch list: a list of alien species that are not yet known to be present in a given area but which are assessed to be likely to be introduced and to have undesirable consequences if they are (cf. **permitted list** and **prohibited list**). *Note: This is sometimes used at a local level to refer to species that need to be monitored for potential future impacts.*

Weed: a plant (regardless of its origin) that has some undesirable impact. It is a purely human (operational) definition. Throughout the book we use 'weed' only to refer to invasive plants that are deemed to have a negative impact, although many of the comments will apply equally to native plants. The IPPC defines a 'pest' as any species, strain, or biotype of plant, animal, or pathogenic agent injurious to plants or plant products, and a quarantine pest as a pest of potential economic importance to the area, endangered thereby and not yet present there, or present but not widely distributed and being officially controlled (FAO 2007).

References

Ackermann, F. & Eden, C. (2011) Strategic management of stakeholders: theory and practice. *Long Range Planning*, **44**, 179–196.

Adams, V. M., & Setterfield, S. A. (2015) Optimal dynamic control of invasions: applying a systematic conservation approach. *Ecological Applications*, **25**, 1131–1141.

Aikio, S., Duncan, R. P., & Hulme, P. E. (2010a) Herbarium records identify the role of long-distance spread in the spatial distribution of alien plants in New Zealand. *Journal of Biogeography*, **37**, 1740–1751.

Aikio, S., Duncan, R. P., & Hulme, P. E. (2010b) Lag-phases in alien plant invasions: separating the facts from the artefacts. *Oikos*, **119**, 370–378.

Alexander, J. M., Kueffer, C., Daehler, C. C., *et al.* (2011) Assembly of nonnative floras along elevational gradients explained by directional ecological filtering. *Proceedings of the National Academy of Sciences of the United States of America*, **108**, 656–661.

Amano, T. & Sutherland, W. J. (2013) Four barriers to the global understanding of biodiversity conservation: wealth, language, geographical location and security. *Proceedings of the Royal Society B: Biological Sciences*, **280**, doi: 10.1098/rspb.2012.2649

Andersen, M. C., Adams, H., Hope, B., & Powell, M. (2004) Risk assessment for invasive species. *Risk Analysis*, **24**, 787–793.

Anderson, L. W. J. (2005) California's reaction to *Caulerpa taxifolia*: a model for invasive species rapid response. *Biological Invasions*, **7**, 1003–1016.

Appleby, A. P. (2005) A history of weed control in the United States and Canada: a sequel. *Weed Science*, **53**, 762–768.

Armstrong, K. F. & Ball, S. L. (2005) DNA barcodes for biosecurity: invasive species identification. *Philosophical Transactions of the Royal Society of London Series B: Biological Sciences*, **360**, 1813–1823.

ATS (2012) Non-Native species online manual. Secretariat for the Antarctic Treaty, www.ats.aq/e/ep_faflo_nns.htm (accessed 27 January 2015).

Auld, B. & Johnson, S. B. (2014) Invasive alien plant management. *CAB Reviews*, **9** (37), 1–12.

Bacon, S. J., Bacher, S., & Aebi, A. (2012) Gaps in border controls are related to quarantine alien insect invasions in Europe. *PLoS ONE*, **7**, e47689, doi: 47610.41371/journal.pone.0047689.

Bailey, S. A., Deneau, M. G., Jean, L., *et al.* (2011) Evaluating efficacy of an environmental policy to prevent biological invasions. *Environmental Science & Technology*, **45**, 2554–2561.

Baker, H. G. (1965) Characteristics and modes of origin of weeds. *The Genetics of Colonizing Species* (eds H. G. Baker & G. L. Stebbins), pp. 147–172. Academic Press, New York.

Barker, K., Taylor, S. L., & Dobson, A. (2013) Interrogating bio-insecurities in biosecurity. *Biosecurity: The Socio-politics of Invasive Species and Infectious Diseases* (eds A. Dobson, K. Barker, & S. L. Taylor), pp. 3–27. Routledge, New York.

Barney, J. N. (2012) Best management practices for bioenergy crops: reducing the invasion risk. Virginia Cooperative Extension, Publication PPWS-8P.

Barney, J. N., Smith, L. L., & Tekiela, D. R. (2014) Using weed risk assessments to parse the weeds from the crops. *Bioenergy and Biological Invasions: Ecological, Agronomic and Policy Perspectives on Minimising Risk* (eds L. D. Quinn, D. P. Matlaga, & J. N. Barney). CABI, Wallingford.

Baxter, P. W. J. & Possingham, H. P. (2011) Optimizing search strategies for invasive pests: learn before you leap. *Journal of Applied Ecology*, **48**, 86–95.

Beaumont, L. J., Gallagher, R. V., Leishman, M. R., Hughes, L., & Downey, P. O. (2014) How can knowledge of the climate niche inform the weed risk assessment process? A case study of *Chrysanthemoides monilifera* in Australia. *Diversity and Distributions*, 20, 613–625.

Bentivegna, D. J., Smeda, R. J., & Wang, C. Z. (2012) Detecting cutleaf teasel (*Dipsacus laciniatus*) along a Missouri highway with hyperspectral imagery. *Invasive Plant Science and Management*, **5**, 155–163.

Benvenuti, S. (2007) Weed seed movement and dispersal strategies in the agricultural environment. *Weed Biology and Management*, **7**, 141–157.

Biosecurity New Zealand (2011) Pest management national plan of action. Ministry of Agriculture and Forestry, New Zealand Government, Wellington.

Biosecurity Queensland (2009) National weed incursion plan preparedness and response guidelines for weed managers. Department of Primary Industries and Fisheries, Queensland, Australia.

Biosecurity Queensland (2013) National Siam weed (*Chromolaena odorata*) eradication program: final report. Department of Agriculture, Fisheries and Forestry, Brisbane.

Blackburn, T. M., Essl, F., Evans, T., *et al.* (2014) A unified classification of alien species based on the magnitude of their environmental impacts. *PLoS Biology*, **12**, e1001850, doi: 1001810.1001371/journal.pbio.1001850.

Blackburn, T. M., Pyšek, P., Bacher, S., *et al.* (2011) A proposed unified framework for biological invasions. *Trends in Ecology & Evolution*, **26**, 333–339.

Bois, S. T., Silander, J. A., & Mehrhoff, L. J. (2011) Invasive plant atlas of New England: the role of citizens in the science of invasive alien species detection. *Bioscience*, **61**, 763–770.

Bourdôt, G. W., Lamoureaux, S. L., Kriticos, D. J., Watt, M. S., & Brown, M. (2010) Current and potential distributions of *Nassella neesiana* (Chilean needle grass) in Australia and New Zealand. *Proceedings of the 17th Australasian Weeds Conference 2010* (ed. S. M. Zydenbos), pp. 424–427. New Zealand Plant Protection Society, Christchurch.

Bourne, L. & Walker, D. H. (2005) Visualising and mapping stakeholder influence. *Management Decision*, **43**, 649–660.

Boy, G. & Witt, A. (2013) *Invasive Alien Plants and their Management in Africa*. CABI, Nairobi.

Brasier, C. M. (2008) The biosecurity threat to the UK and global environment from international trade in plants. *Plant Pathology*, **57**, 792–808.

Broennimann, O. & Guisan, A. (2008) Predicting current and future biological invasions: both native and invaded ranges matter. *Biology Letters*, **4**, 585–589.

Brooks, M. L. & Klinger, R. C. (2009) Practical considerations for early detection monitoring of plant invasions. *Management of Invasive Weeds* (ed. Inderjit), pp. 9–33. Springer, New York.

Brooks, S. J. (2012) Ecology and control of national weed eradication targets. *Technical Highlights: Invasive Plant and Animal Research 2010–11*, pp. 39–41. Department of Primary Industries and Fisheries, Brisbane.

Brooks, S. J., Panetta, F. D., & Galway, K. E. (2008) Progress towards the eradication of Mikania Vine (*Mikania micrantha*) and Limnocharis (*Limnocharis flava*) in Northern Australia. *Invasive Plant Science and Management*, **1**, 296–303.

Brooks, S. J., Panetta, F. D., & Sydes, T. A. (2009) Progress towards the eradication of three melastome shrub species from northern Australian rainforests. *Plant Protection Quarterly*, **24**, 71–78.

Brown, R. B. & Noble, S. D. (2005) Site-specific weed management: sensing requirements – what do we need to see? *Weed Science*, **53**, 252–258.

Brunel, S. (2014) How to communicate on pests and invasive alien plants? Conclusions of the EPPO/CoE/IUCN- ISSG/DGAV/UC/ESAC Workshop. A workshop to bridge the gap in between disciplines. *Bulletin OEPP/EPPO Bulletin*, **44**, 205–211.

Bryson, J. M. & Roering, W. D. (1988) Initiation of strategic planning by governments. *Public Administration Review*, **48**, 995–1004.

Buckley, Y. M., Bolker, B. M., & Rees, M. (2007) Disturbance, invasion and reinvasion: managing the weed-shaped hole in disturbed ecosystems. *Ecology Letters*, **10**, 809–817.

Buckley, Y. M., Brockerhoff, E., Langer, L., et al. (2005) Slowing down a pine invasion despite uncertainty in demography and dispersal. *Journal of Applied Ecology*, **42**, 1020–1030.

Buddenhagen, C. E. & Tye, A. (2015) Lessons from successful plant eradications in Galapagos: commitment is crucial. *Biological Invasions*, doi: 10.1007/s10530-10015-10919-y.

Bufford, J. L. & Daehler, C. C. (2014) Sterility and lack of pollinator services explain reproductive failure in non-invasive ornamental plants. *Diversity and Distributions*, **20**, 975–985.

Bullock, J. M., Pywell, R. F., & Coulson-Phillips, S. J. (2008) Managing plant population spread: prediction and analysis using a simple model. *Ecological Applications*, **18**, 945–953.

Burgiel, S., Foote, G., Oreliana, M., & Perrault, A. (2006) Invasive alien species and trade: integrating prevention measures and international trade rules. Center for International Environmental Law.

Burgman, M. A., McCarthy, M. A., Robinson, A., et al. (2013) Improving decisions for invasive species management: reformulation and extensions of the Panetta–Lawes eradication graph. *Diversity and Distributions*, **19**, 603–607.

Burnette, R., Hess, J. E., Kozlovac, J. P., & Richmond, J. Y. (2013) Defining biosecurity and related concepts. *Biosecurity: Understanding, Assessing, and Preventing Threat* (ed. R. Burnette), pp. 3–14. John Wiley & Sons, Hoboken, NJ.

Burt, J. W., Muir, A. A., Piovia-Scott, J., Veblen, K. E., Chang, A. L., Grossman, J. D., & Weiskel, H. W. (2007) Preventing horticultural introductions of invasive plants: potential efficacy of voluntary initiatives. *Biological Invasions*, **9**, 909–923.

Cacho, O. (2004) When is it optimal to eradicate a weed invasion? *Proceedings of the 14th Australian Weeds Conference* (eds B. M. Sindel & S. B. Johnson), pp. 49–54. Weeds Society of New South Wales, Sydney.

Cacho, O. J. & Hester, S. M. (2011) Deriving efficient frontiers for effort allocation in the management of invasive species. *Australian Journal of Agricultural and Resource Economics*, **55**, 72–89.

Cacho, O. J., Hester, S., & Spring, D. (2007) Applying search theory to determine the feasibility of eradicating an invasive population in natural environments. *Australian Journal of Agricultural and Resource Economics*, **51**, 425–443.

Cacho, O. J. & Pheloung, P. (2007) *WeedSearch: Weed Eradication Feasibility Analysis, Software Manual*. Co-operative Research Centre for Australian Weed Management, Project 1.2.8, August 2007, University of New England, Australia. www-personal.une.edu.au/~ocacho/weedsearch.htm.

Cacho, O. J., Spring, D., Hester, S., & MacNally, R. (2010) Allocating surveillance effort in the management of invasive species: a spatially-explicit model. *Environmental Modelling & Software*, **25**, 444–454.

Cacho, O. J., Spring, D., Pheloung, P., & Hester, S. (2006) Evaluating the feasibility of eradicating an invasion. *Biological Invasions*, **8**, 903–917.

Cacho, O. J., Wise, R. M., Hester, S. M., & Sinden, J. A. (2008) Bioeconomic modeling for control of weeds in natural environments. *Ecological Economics*, **65**, 559–568.

Caley, P., Groves, R. H., & Barker, R. (2008) Estimating the invasion success of introduced plants. *Diversity and Distributions*, **14**, 196–203.

Campbell, F. T. (2001) The science of risk assessment for phytosanitary regulation and the impact of changing trade regulations. *Bioscience*, **51**, 148–153.

Caplat, P., Hui, C., Maxwell, B., & Peltzer, D. (2014) Cross-scale management strategies for optimal control of trees invading from source plantations. *Biological Invasions*, **16**, 677–690.

Castro-Díez, P., Godoy, O., Alonso, A., Gallardo, A., & Saldaña, A. (2014) What explains variation in the impacts of exotic plant invasions on the nitrogen cycle? A meta-analysis. *Ecology Letters*, **17**, 1–12.

Castro-Díez, P., Godoy, O., Saldaña, A., & Richardson, D. M. (2011) Predicting invasiveness of Australian acacias on the basis of their native climatic affinities, life history traits and human use. *Diversity and Distributions*, **17**, 934–945.

Chadès, I., Martin, T. G., Nicol, S., et al. (2011) General rules for managing and surveying networks of pests, diseases, and endangered species. *Proceedings of the National Academy of Sciences of the United States of America*, **108**, 8323–8328.

Chaffee, E. E. (1985) Three models of strategy. *Academy of Management Review*, **10**, 89–98.

Champion, P. D. & Clayton, J. S. (2001) A weed risk assessment model for aquatic plants in New Zealand. *Weed Risk Assessment* (eds R. H. Groves, F. D. Panetta, & J. G. Virtue), pp. 194–202. CSIRO, Melbourne.

Champion, P. D., Clayton, J. S., Petroechevsky, A., & Newfield, M. (2010) Using the New Zealand aquatic risk assessment model to manage potential weeds in the aquarium/pond plant trade. *Plant Protection Quarterly*, **25**, 49–51.

Chen, G., Kery, M., Zhang, J., & Ma, K. (2009) Factors affecting detection probability in plant distribution studies. *Journal of Ecology*, **97**, 1383–1389.

Chown, S. L., Gremmen, N. J. M., & Gaston, K. J. (1998) Ecological biogeography of Southern Ocean Islands: species–area relationships, human impacts, and conservation. *The American Naturalist*, **52**, 562–575.

Chown, S. L., Huiskes, A. H. L., Gremmen, N. J. M., *et al.* (2012) Continent-wide risk assessment for the establishment of nonindigenous species in Antarctica. *Proceedings of the National Academy of Sciences of the United States of America*, **109**, 4938–4943.

Colautti, R. I., Grigorovich, I. A., & MacIsaac, H. J. (2006) Propagule pressure: a null model for biological invasions. *Biological Invasions*, **8**, 1023–1037.

Colwell, R. K. & Rangel, T. F. (2009) Hutchinson's duality: the once and future niche. *Proceedings of the National Academy of Sciences of the United States of America*, **106**, 19651–19658.

COMNAP/SCAR (2010) Non-native species checklist for supply managers. Committee for Managers of National Antarctic Programs. The Scientific Committee for Antarctic Research, www.comnap.aq/SitePages/checklists.aspx (accessed 27 January 2015).

Cook, C. N., Inayatullah, S., Burgman, M. A., Sutherland, W. J., & Wintle, B. A. (2014) Strategic foresight: how planning for the unpredictable can improve environmental decision-making. *Trends in Ecology & Evolution*, **29**, 521–541.

Cook, G. D. & Dias, L. (2006) It was no accident: deliberate plant introductions by Australian government agencies during the 20th century. *Australian Journal of Botany*, **54**, 601–625.

Corn, M. L. & Johnson, R. (2013) Invasive species: major laws and the role of selected federal agencies. Congressional Research Service Report 7–5700.

Cousens, R. & Mortimer, M. (1995) *Dynamics of Weed Populations*, 1st edn. Cambridge University Press, Cambridge.

Coutts, S., van Klinken, R. D., Yokomizo, H., & Buckley, Y. M. (2011) What are the key drivers of spread in invasive plants: dispersal, demography, or landscape – and how can we use this knowledge to aid management? *Biological Invasions*, **13**, 1649–1661.

Crall, A. W., Newman, G. J., Stohlgren, T. J., *et al.* (2011) Assessing citizen science data quality: an invasive species case study. *Conservation Letters*, **4**, 433–442.

Crooks, J. A. (2005) Lag times and exotic species: the ecology and management of biological invasions in slow-motion. *Ecoscience*, **12**, 316–329.

Crooks, J. A. (2011) Lag times. *Encyclopedia of Biological Invasions* (eds D. Simberloff & M. Rejmánek), pp. 404–408. University of California Press, Berkeley and Los Angeles.

Cunningham, D. C., Woldendorp, G., Burgess, M. B., & Barry, S. C. (2003) *Prioritising Sleeper Weeds for Eradication: Selection of Species Based on Potential Impacts on Agriculture and Feasibility of Eradication*. Bureau of Resource Sciences, Canberra.

Daehler, C. C. (1998) The taxonomic distribution of invasive angiosperm plants: ecological insights and comparison to agricultural weeds. *Biological Conservation*, **84**, 167–180.

Daehler, C. C. (2009) Short lag times for invasive tropical plants: evidence from experimental plantings in Hawai'i. *PLoS ONE*, **4**, e4462, doi: 4410.1371/journal.pone.0004462.

Daehler, C. C., Denslow, J. E., Ansari, S., & Kuo, H.-C. (2004) A risk-assessment system for screening out invasive pest plants from Hawaii and other Pacific Islands. *Conservation Biology*, **18**, 360–369.

Daehler, C. C. & Virtue, J. G. (2010) Likelihood and consequence: reframing the Australian Weed Risk Assessment to reflect a standard model of risk. *Plant Protection Quarterly*, **25**, 51–55.

Dawson, W., Burslem, D. F. R. P., & Hulme, P. E. (2009) Factors explaining alien plant invasion success in a tropical ecosystem differ at each stage of invasion. *Journal of Ecology*, **97**, 657–665.

de Villiers, M. S., Cooper, J., Carmichael, N., *et al.* (2005) Conservation management at southern ocean islands: towards the development of best-practice guidelines. *Polarforschung*, **75**, 113–131.

DEFRA (2003) *Review of Non-native Species Policy*. Department for Environment, Food and Rural Affairs, London.

del-Val, E., Balvanera, P., Castellarini, F., *et al.* (2015) Identifying areas of high invasion risk: a general model and an application to Mexico. *Revista Mexicana De Biodiversidad*, **86**, 208–216.

Delaney, D. G., Sperling, C. D., Adams, C. S., & Leung, B. (2008) Marine invasive species: validation of citizen science and implications for national monitoring networks. *Biological Invasions*, **10**, 117–128.

DeLoach, C. J., Carruthers, R. I., Lovich, J. E., Dudley, T. L., & Smith, S. D. (2000) Ecological interactions in the biological control of saltcedar (*Tamarix* spp.) in the United States: toward a new understanding. *Proceedings of the X International Symposium on Biological Control of Weeds* (ed. N. R. Spencer), pp. 819–873. Montana State University, USDA ARS, Bozeman, MT.

Department of Environmental Affairs (2014) National Environmental Management: Biodiversity Act 2004 (Act No, 10 of 2004) draft alien and invasive species lists, 2014 & draft alien and invasive species regulations, 2014. (ed. DEA), pp. 3–92. Government Gazette, Pretoria.

Devictor, V., Whittaker, R. J., & Beltrame, C. (2010) Beyond scarcity: citizen science programmes as useful tools for conservation biogeography. *Diversity and Distributions*, **16**, 354–362.

Devorshak, C. (2012) *Plant Pest Risk Analysis: Concepts and Application*. CABI, Wallingford.

Diez, J. M., Hulme, P. E., & Duncan, R. P. (2012) Using prior information to build probabilistic invasive species risk assessments. *Biological Invasions*, **14**, 681–691.

DiTomaso, J. M. (1998) Impact, biology, and ecology of saltcedar (*Tamarix* spp.) in the Southwestern United States. *Weed Technology*, **12**, 326–336.

Dobson, A., Barker, K., & Taylor, S. L. (2013) *Biosecurity: The Socio-politics of Invasive Species and Infectious Diseases*. Routledge, New York.

Dodd, A. J., Ainsworth, N., Burgman, M. A., & McCarthy, M. A. (2015) Plant extirpation at the site scale: implications for eradication programmes. *Diversity and Distributions*, **21**, 151–162.

Dodd, A. J., McCarthy, M. A., Ainsworth, N., & Burgman, M. A. (2016) Identifying hotspots of alien plant naturalisation in Australia: approaches and predictions. *Biological Invasions*, **18**, 631–645.

Donaldson, J. E., Hui, C., Richardson, D. M., et al. (2014) Invasion trajectory of alien trees: the role of introduction pathway and planting history. *Global Change Biology*, **20**, 1527–1537.

Donaldson, J. E., Richardson, D. M., & Wilson, J. R. U. (2014) Scale–area curves identify artefacts of human use in the spatial structure of an invasive tree. *Biological Invasions*, **16**, 553–563.

Donlan, C. J. & Wilcox, C. (2007) Complexities of costing eradications. *Animal Conservation*, **10**, 154–156.

Dorazio, R. M. (2014) Accounting for imperfect detection and survey bias in statistical analysis of presence-only data. *Global Ecology and Biogeography*, **23**, 1472–1484.

Drenovsky, R. E., Grewell, B. J., D'Antonio, C. M., et al. (2012) A functional trait perspective on plant invasion. *Annals of Botany*, **13**, doi: 10.1093/aob/mcs1100.

Drew, J., Anderson, N., & Andow, D. (2010) Conundrums of a complex vector for invasive species control: a detailed examination of the horticultural industry. *Biological Invasions*, **12**, 2837–2851.

Driscoll, D. A., Catford, J. A., Barney, J. N., et al. (2014) New pasture plants intensify invasive species risk. *Proceedings of the National Academy of Sciences of the United States of America*, **111**, 16622–16627.

Ebbels, D. L. (2003) *Principles of Plant Health and Quarantine*. CABI, Wallingford.

Edwards, P. K. & Leung, B. (2009) Re-evaluating eradication of nuisance species: invasion of the tunicate, *Ciona intestinalis*. *Frontiers in Ecology and the Environment*, **7**, 326–332.

Elith, J. (in press) Predicting distributions of invasive species. *Risk-Based Decisions for Biological Threats* (eds T. R. Walshe, A. Robinson, M. Nunn, & M. A. Burgman). Cambridge University Press, Cambridge.

Elith, J., Kearney, M., & Phillips, S. (2010) The art of modelling range-shifting species. *Methods in Ecology and Evolution*, **1**, 330–342.

Epanchin-Niell, R. S. & Hastings, A. (2010) Controlling established invaders: integrating economics and spread dynamics to determine optimal management. *Ecology Letters*, **13**, 528–541.

EPPO (2011) PM 5/3 (5) Guidelines on pest risk analysis: decision-support scheme for quarantine pests. http://archives.eppo.int/EPPOStandards/pra.htm.

EPPO (2012) PM 5/6 EPPO prioritization process for invasive alien plants. *Bulletin OEPP/EPPO Bulletin*, **42**, 463–474, http://archives.eppo.int/EPPOStandards/pra.htm.

Esler, K. J., Pozesky, H., Sharma, G. P., & McGeoch, M. (2010) How wide is the 'knowing–doing' gap in invasion biology? *Biological Invasions*, **12**, 4065–4075.

Essl, F., Dullinger, S., Rabitsch, W., et al. (2011) Socioeconomic legacy yields an invasion debt. *Proceedings of the National Academy of Sciences of the United States of America*, **108**, 203–207.

Evans, C. L. (2002) *War on Weeds in the Prairie West: An Environmental History.* University of Calgary Press, Calgary.
FAO (2007) *FAO Biosecurity Toolkit.* Food and Agricultural Organization of the United Nations, Rome.
FAO (2012) *Protecting the World's Plant Resources From Pests: An International Framework for Cooperation.* Food and Agriculture Organization, Rome. www.ippc.int/sites/default/files/mediakit/IPPCOverviewBrochure2012−03-en.pdf.
Faulkner, K. T., Robertson, M. P., Rouget, M., & Wilson, J. R. U. (2014) A simple, rapid methodology for developing invasive species watch lists. *Biological Conservation,* **179**, 25–32.
Fenner, F., Henderson, D. A., Arita, I., Ježek, Z., & Ladnyi, I. D. (1988) Lessons and benefits. *Smallpox and its Eradication,* pp. 1345–1370. World Health Organization, Geneva.
Finnoff, D., Shogren, J. F., Leung, B., & Lodge, D. (2007) Take a risk: preferring prevention over control of biological invaders. *Ecological Economics,* **62**, 216–222.
Fithian, W., Elith, J., Hastie, T., & Keith, D. A. (2015) Bias correction in species distribution models: pooling survey and collection data for multiple species. *Methods in Ecology and Evolution,* **6**, 424–438.
Fletcher, C. S., Westcott, D. A., Murphy, H. T., Grice, A. C., & Clarkson, J. R. (2015) Managing breaches of containment and eradication of invasive plant populations. *Journal of Applied Ecology,* **52**, 59–68.
Fox, J. C., Buckley, Y. M., Panetta, F. D., Bourgoin, J., & Pullar, D. (2009) Surveillance protocols for management of invasive plants: modelling Chilean needle grass (*Nassella neesiana*) in Australia. *Diversity and Distributions,* **15**, 577–589.
Frenot, Y., Chown, S. L., Whinam, J., et al. (2005) Biological invasions in the Antarctic: extent, impacts and implications. *Biological Reviews,* **80**, 45–72.
Gaertner, M., Holmes, P. M., & Richardson, D. M. (2012) Biological invasions, resilience and restoration. *Restoration Ecology: The New Frontier* (eds J. Andel & J. Aronson), pp. 265–280. Wiley-Blackwell, Oxford.
Gage, E. A. & Cooper, D. J. (2005) Patterns of willow seed dispersal, seed entrapment, and seedling establishment in a heavily browsed montane riparian ecosystem. *Canadian Journal of Botany/Revue Canadienne De Botanique,* **83**, 678–687.
García, G. & Gardener, M. R. (2012) Evaluación de proyectos de control de plantas transformadores y reforestación de sitios de alta valor en Galápagos. Unpublished report, Galapagos National Park, Puerto Ayora, Ecuador.
Gardener, M. R., Atkinson, R., & Rentería, J. L. (2010) Eradications and people: lessons from the plant eradication program in Galapagos. *Restoration Ecology,* **18**, 20–29.
Gardener, M. R., Trueman, M., Buddenhagen, C., et al. (2013) A pragmatic approach to the management of plant invasions in Galapagos. *Plant Invasions in Protected Areas: Patterns, Problems and Challenges* (eds L. C. Foxcroft, P. Pyšek, D. M. Richardson, & P. Genovesi), pp. 349–374. Springer, New York.
Garner, B. A. (2009) *Black's Law Dictionary.* 9th edn. West Publishing Company, St Paul, MN.
Garrard, G. E., Bekessy, S. A., McCarthy, M. A., & Wintle, B. A. (2008) When have we looked hard enough? A novel method for setting minimum survey effort protocols for flora surveys. *Austral Ecology,* **33**, 986–998.

Geerts, S., Mashele, B., Visser, V., & Wilson, J. R. U. (in review) Right place in the wrong way: lack of human assisted dispersal means *Pueraria montana* var. lobata (kudzu vine) could still be eradicated from South Africa.

Geerts, S., Moodley, D., Gaertner, M., *et al.* (2013) The absence of fire can cause a lag phase: the invasion dynamics of *Banksia ericifolia* (Proteaceae). *Austral Ecology*, **38**, 931–941.

Gerlach, J. D., Bushman, B. S., McKay, J. K., & Meimberg, H. (2009) Taxonomic confusion permits the unchecked invasion of vernal pools in California by low mannagrass (*Glyceria declinata*). *Invasive Plant Science and Management*, **2**, 92–97.

Giessow, J., Casanova, J., Leclerc, R., *et al.* (2011) *Arundo donax* (giant reed): distribution and impact report. State Water Resources Control Board, Agreement No. 06-374-559-0.

Giljohann, K. M., Hauser, C. E., Williams, N. S. G., & Moore, J. L. (2011) Optimizing invasive species control across space: willow invasion management in the Australian Alps. *Journal of Applied Ecology*, **48**, 1286–1294.

Gobster, P. H. (2011) Factors affecting people's response to invasive species management. *Invasive and Introduced Plants and Animals: Human Perceptions, Attitudes and Approaches to Management* (eds I. Rotheram & R. Lambert), pp. 249–263. Earthscan, London.

Goodwin, K. M., Engel, R. E., & Weaver, D. K. (2010) Trained dogs outperform human surveyors in the detection of rare spotted knapweed (*Centaurea stoebe*). *Invasive Plant Science and Management*, **3**, 113–121.

Gordon, D. R., Mitterdorfer, B., Pheloung, P. C., *et al.* (2010) Guidance for addressing the Australian Weed Risk Assessment questions. *Plant Protection Quarterly*, **25**, 56–74.

Gordon, D. R., Onderdonk, D. A., Fox, A. M., & Stocker, R. K. (2008) Consistent accuracy of the Australian weed risk assessment system across varied geographies. *Diversity and Distributions*, **14**, 234–242.

Gravuer, K., Sullivan, J. J., Williams, P. A., & Duncan, R. P. (2008) Strong human association with plant invasion success for *Trifolium* introductions to New Zealand. *Proceedings of the National Academy of Sciences of the United States of America*, **105**, 6344–6349.

Grevstad, F. S. (2005) Simulating control strategies for a spatially structured weed invasion: *Spartina alterniflora* (Loisel) in Pacific Coast estuaries. *Biological Invasions*, **7**, 665–677.

Grice, T. (2009) Principles of containment and control of invasive species. *Invasive Species Management: A Handbook of Techniques* (eds M. N. Clout & P. A. Williams), pp. 61–76. Oxford University Press, Oxford.

Groves, R. H. (1991) The biogeography of Mediterranean plant invasions. *Biogeography of Mediterranean Invasions* (eds R. H. Groves & F. Di Castri), pp. 427–438. Cambridge University Press, Cambridge.

Gruszczynski, L. (2006) The role of science in risk regulation under the SPS agreement. EUI Working Papers Law 2006/03, European Institute University Department of Law.

Guillera-Arroita, G., Lahoz-Monfort, J. J., Elith, J., *et al.* (2015) Is my species distribution model fit for purpose? Matching data and models to applications. *Global Ecology and Biogeography*, **24**, 276–292.

Halford, M., Heemers, L., van Wesemael, D., et al. (2014) The voluntary code of conduct on invasive alien plants in Belgium: results and lessons learned from the AlterIAS LIFE+ project. *EPPO Bulletin*, **44**, 1–11.

Harris, S. & Timmins, S. M. (2009) Estimating the benefit of early control of all newly naturalised plants. Science for Conservation, Department of Conservation, Wellington, New Zealand.

Hauser, C. E., Giljohann, K. M., Rigby, M., et al. (2016) Practicable methods for delimiting a plant invasion. *Diversity and Distributions*, **22**, 136–147.

Hauser, C. E. & McCarthy, M. A. (2009) Streamlining 'search and destroy': cost-effective surveillance for invasive species management. *Ecology Letters*, **12**, 683–692.

Hauser, C. E., Moore, J. L., Garrard, G. E., & McCarthy, M. A. (2012) Designing a detection experiment: tricks and trade-offs. *Proceedings of the 18th Australasian Weeds Conference* (ed. V. Eldershaw), pp. 267–272, www.caws.org.au/awc/2012/awc201212671.pdf.

Hawkins, C. L., Bacher, S., Essl, F., et al. (2015) Framework and guidelines for implementing the proposed IUCN Environmental Impact Classification for Alien Taxa (EICAT). *Diversity and Distributions*, **21**, 1360–1363.

Hedley, J. (2004) The International Plant Protection Convention and invasive species. *Harmful Invasive Species: Legal Response* (eds M. L. Miller & R. N. Fabian), pp. 185–201. Environmental Law Institute, Washington, DC.

Henderson, L. (2007) Invasive, naturalized and casual alien plants in southern Africa: a summary based on the southern African Plant Invaders Atlas (SAPIA). *Bothalia*, **37**, 215–248.

Henne, D. & Lindgren, C. J. (2005) An integrated management strategy for the control of Purple Loosestrife (*Lythrum salicaria*) L. (Lythraceae) in the Netley-Libau Marsh, southern Manitoba. *Biological Control*, **32**, 319–325.

Hester, S. M., Brooks, S. J., Cacho, O. J., & Panetta, F. D. (2010) Applying a simulation model to the management of an infestation of *Miconia calvescens* in the wet tropics of Australia. *Weed Research*, **50**, 269–279.

Hester, S. M., Cacho, O. J., Panetta, F. D., & Hauser, C. E. (2013) Economic aspects of post-border weed risk management. *Diversity and Distributions*, **19**, 580–589.

Hester, S. M. & Cacho, O. J. (in review) The contribution of passive surveillance to invasive species management.

Hewitt, C. L., Everett, R. A., & Parker, N. (2009) Examples of current international, regional and national regulatory frameworks for preventing and managing marine bioinvasions. *Biological Invasions in Marine Ecosystems* (eds G. Rilov & J. A. Cook), pp. 335–352. Springer-Verlag, Berlin.

Higgins, S. I., Nathan, R., & Cain, M. L. (2003) Are long-distance dispersal events in plants usually caused by nonstandard means of dispersal? *Ecology*, **84**, 1945–1956.

Higgins, S. I., Richardson, D. M., & Cowling, R. M. (2000) Using a dynamic landscape model for planning the management of alien plant invasions. *Ecological Applications*, **10**, 1833–1848.

Hill, R. L., Gourlay, A. H., & Barker, R. J. (2001) Survival of *Ulex europaeus* seeds in the soil at three sites in New Zealand. *New Zealand Journal of Botany*, **39**, 235–244.

Himmelman, A. T. (1992) Communities working collaboratively for a change. Humphrey Institute for Public Affairs, University of Minneapolis, Minneapolis.

Hirzel, A. H. & Lay, G. L. (2008) Habitat suitability modelling and niche theory. *Journal of Applied Ecology*, **45**, 1372–1381.

Hobbs, R. J., Arico, S., Aronson, J., et al. (2006) Novel ecosystems: theoretical and management aspects of the new ecological world order. *Global Ecology and Biogeography*, **15**, 1–7.

Hobbs, R. J. & Humphries, S. E. (1995) An integrated approach to the ecology and management of plant invasions. *Conservation Biology*, **9**, 761–770.

Hodur, N. M., Leistritz, F. L., & Bangsund, D. A. (2006) Evaluation of TEAM leafy spurge project. *Rangeland Ecology and Management*, **59**, 483–493.

Hohmann, M. G., Just, M. G., Frank, P. J., Wall, W. A., & Gray, J. B. (2013) Prioritising invasive plant management with multi-criteria decision analysis. *Invasive Plant Science and Management*, **6**, 339–351.

Howell, C. J. (2012) Progress toward environmental weed eradication in New Zealand. *Invasive Plant Science and Management*, **5**, 249–258.

Huang, D. C., Zhang, R. Z., Kim, K. C., & Suarez, A. V. (2012) Spatial pattern and determinants of the first detection locations of invasive alien species in mainland China. *PLoS ONE*, **7**, e31734, doi: 31710.31371/journal.pone.0031734.

Hughes, K. A. & Convey, P. (2010) The protection of Antarctic terrestrial ecosystems from inter- and intra-continental transfer of non-indigenous species by human activities: a review of current systems and practices. *Global Environmental Change: Human and Policy Dimensions*, **20**, 96–112.

Hughes, K. A., Pertierra, L. R., Molina Montenegro, M. A., & Convey, P. (2015) Biological invasions in terrestrial Antarctica: what is the current status and can we respond? *Biodiversity Conservation*, **24**, 1031–1055.

Hui, C., Richardson, D. M., Robertson, M. P., Wilson, J. R. U., & Yates, C. Y. (2011) Macroecology meets invasion ecology: linking native distribution of Australian acacias to invasiveness. *Diversity and Distributions*, **17**, 872–883.

Huiskes, A. H. L., Gremmen, N. J. M., Bergstrom, D. M., et al. (2014) Aliens in Antarctica: assessing transfer of plant propagules by human visitors to reduce invasion risk. *Biological Conservation*, **171**, 278–284.

Hulme, P. E. (2006) Beyond control: wider implications for the management of biological invasions. *Journal of Applied Ecology*, **43**, 835–847.

Hulme, P. E. (2011) Addressing the threat to biodiversity from botanic gardens. *Trends in Ecology & Evolution*, **26**, 168–174.

Hulme, P. E. (2012) Weed risk assessment: a way forward or a waste of time? *Journal of Applied Ecology*, **49**, 10–19.

Hulme, P. E. (2014) An introduction to plant biosecurity: past, present and future. *The Handbook of Plant Biosecurity* (eds G. Gordh & S. McKirdy), pp. 1–25. Springer, Dordrecht.

Hulme, P. E., Bacher, S., Kenis, M., et al. (2008) Grasping at the routes of biological invasions: a framework for integrating pathways into policy. *Journal of Applied Ecology*, **45**, 403–414.

Hulme, P. E., Pyšek, P., Jarošík, V., Pergl, J., Schaffner, U., & Vilà, M. (2013) Bias and error in understanding plant invasion impacts. *Trends in Ecology & Evolution*, **28**, 212–218.

Hulme, P. E., Pyšek, P., Nentwig, W., & Vilá, M. (2009) Will threat of biological invasions unite the European Union? *Science*, **324**, 40–41.

IPPC (2005) Identification of risks and management of invasive alien species using the IPPC framework. International Plant Protection Convention Secretariat,

FAO, Workshop in Braunschweig, Germany, 22–26 September 2003. Rome, Italy.

IPPC-FAO (2003) *International Standards for Phytosanitary Measures (ISPM) 19: Guidelines on Lists of Regulated Pests*. Food and Agricultural Organization of the United Nations, Rome. www.ippc.int/static/media/files/publications/en/ 1323945631_ISPM_1323945619_1323942003_En_1323942011–1323945611 –1323945629_Refor.pdf.

IPPC-FAO (2006) *International Standards for Phytosanitary Measures (ISPM) 9: Guidelines for Pest Eradication Programs*. pp. 95–102, Food and Agricultural Organization of the United Nations, Rome. ftp://ftp.fao.org/docrep/fao/009/a0450e/ a0450e0400.pdf.

IPPC-FAO (2007) *International Standards for Phytosanitary Measures (ISPM) 2: Framework for Pest Risk Analysis*. Food and Agricultural Organization of the United Nations, Rome.

IPPC-FAO (2013a) *International Standards for Phytosanitary Measures (ISPM) 5: Glossary of Phytosanitary Terms*. Food and Agricultural Organization of the United Nations, Rome. www.ippc.int/sites/default/files/documents/20140214/ispm_ 20140205_en_20142014–20140202–20140214cpm-20140218_201402141055 –201402141559.201402141025%201402141020KB.pdf.

IPPC-FAO (2013b) *International Standards for Phytosanitary Measures (ISPM) 11: Pest Risk Analysis for Quarantine Pests*. Food and Agricultural Organization of the United Nations, Rome.

Jacobs, L. E. O., Richardson, D. M., & Wilson, J. R. U. (2014) *Melaleuca parvistaminea* Byrnes (Myrtaceae) in South Africa: invasion risk and feasibility of eradication. *South African Journal of Botany*, 94, 24–32.

Jenkins, P. T. (1996) Free trade and exotic species introductions. *Conservation Biology*, **10**, 300–302.

Jiménez-Valverde, A., Peterson, A. T., Soberon, J., et al. (2011) Use of niche models in invasive species risk assessments. *Biological Invasions*, **13**, 2785–2797.

Kahneman, D. (2011) *Thinking, Fast and Slow*. Penguin Books, London.

Kaplan, H., van Niekerk, A., Le Roux, J. J., Richardson, D. M., & Wilson, J. R. U. (2014) Incorporating risk mapping at multiple spatial scales into eradication management plans. *Biological Invasions*, **16**, 691–703.

Kay, S. H. & Hoyle, S. T. (2001) Mail order, the Internet, and invasive aquatic weeds. *Journal of Aquatic Plant Management*, **39**, 88–91.

Kearney, M. R., Isaac, A. P., & Porter, W. P. (2014) microclim: Global estimates of hourly microclimate based on long-term monthly climate averages. *Scientific Data*, **1**, doi: 10.1038/sdata.2014.6.

Kearney, M. & Porter, W. (2009) Mechanistic niche modelling: combining physiological and spatial data to predict species' ranges. *Ecology Letters*, **12**, 334–350.

Kehlenbeck, H., Robinet, C., van der Werf, W., et al. (2012) Modelling and mapping spread in pest risk analysis: a generic approach. *EPPO Bulletin*, **42**, 74–80.

Keller, R. P., Frang, K., & Lodge, D. M. (2008) Preventing the spread of invasive species: economic benefits of intervention guided by ecological predictions. *Conservation Biology*, **22**, 80–88.

Kolar, C. S. & Lodge, D. M. (2001) Progress in invasion biology: predicting invaders. *Trends in Ecology & Evolution*, **16**, 199–204.

Köppen, W. P. (1936) Das Geographische System der Klimate. *Handbuch der Klimatologie* (eds W. Köppen & G. C. Geiger), pp. 1–44. Gebrüder Bornträger, Berlin.

Kowarik, I. (1995) Time lags in biological invasions with regard to the success and failure of alien species. *Plant Invasions: General Aspects and Special Problems* (eds P. Pyšek, K. Prach, M. Rejmánek, & M. Wade), pp. 15–38. SPB Academic Publishing, Amsterdam.

Kriticos, D. J., Phillips, C. B., & Suckling, D. M. (2005) Improving border biosecurity: potential economic benefits to New Zealand. *New Zealand Plant Protection*, **58**, 1–6.

Křivánek, M. & Pyšek, P. (2006) Predicting invasions by woody species in a temperate zone: a test of three risk assessment schemes in the Czech Republic (Central Europe). *Diversity and Distributions*, **12**, 319–327.

Kumschick, S., Bacher, S., Dawson, W., et al. (2012) A conceptual framework for prioritization of invasive alien species for management according to their impact. *Neobiota*, **15**, 69–100.

Kumschick, S., & Richardson, D. M. (2013) Species-based risk assessments for biological invasions: advances and challenges. *Diversity and Distributions*, **19**, 1095–1105.

Lambdon, P. W., Pyšek, P., Basnou, C., et al. (2008) Alien flora of Europe: species diversity, temporal trends, geographical patterns and research needs. *Preslia*, **80**, 101–149.

Le Roux, J. J., & Wieczorek, A. M. (2009) Molecular systematics and population genetics of biological invasions: towards a better understanding of invasive species management. *Annals of Applied Biology*, **154**, 1–17.

Lee, J. E. & Chown, S. L. (2009) Breaching the dispersal barrier to invasion: quantification and management. *Ecological Applications*, **19**, 1944–1959.

Leung, B., Cacho, O. J., & Spring, D. (2010) Searching for non-indigenous species: rapidly delimiting the invasion boundary. *Diversity and Distributions*, **16**, 451–460.

Leung, B., Roura-Pascual, N., Bacher, S., et al. (2012) TEASIng apart alien species risk assessments: a framework for best practices. *Ecology Letters*, **15**, 1475–1493.

Leung, B., Roura-Pascual, N., Bacher, S., et al. (2013) Addressing a critique of the TEASI framework for invasive species risk assessment. *Ecology Letters*, **16**, 1415–1416.

Lindgren, C., Pearce, C., & Allison, K. (2010) The biology of invasive alien plants in Canada: 11. *Tamarix ramosissima* Ledeb., *T. chinensis* Lour. and hybrids. *Canadian Journal of Plant Science*, **90**, 111–124.

Lindgren, C. J. (2000) Performance of a biological control agent, *Galerucella calmariensis* (Coleoptera: Chrysomelidae) on Purple Loosestrife *Lythrum salicaria* L. in southern Manitoba (1993–1998). *Proceedings of the X International Symposium on Biological Control of Weeds* (ed. N. R. Spencer), pp. 367–382. Montana State University, USDA ARS, Bozeman, MT.

Lindgren, C. J. (2002) Manitoba Purple Loosestrife Project: partnerships and initiatives in the control of an invasive alien species. *Alien Invaders in Canada's Waters, Wetlands, and Forests* (eds R. Claudi, P. Nantel, & E. Muckle-Jeffs), pp. 259–267. Canadian Forest Service, Natural Resources Canada, Ottawa.

Lindgren, C. J. (2003) A brief history of Purple Loosestrife, *Lythrum salicaria*, in Manitoba and its status in 2001. *Canadian Field Naturalist*, **117**, 100–109.

Lindgren, C. J. (2012) Biosecurity policy and the use of geospatial predictive tools to address invasive plants: updating the risk analysis toolbox. *Risk Analysis*, **32**, 9–15.

Lindgren, C. J. & Clay, R. (1993) Fertility of 'Morden Pink' *Lythrum virgatum* L. transplanted into wild stands of *L. salicaria* in Manitoba. *HortScience*, **28**, 954.

Lockwood, J. L., Cassey, P., & Blackburn, T. (2005) The role of propagule pressure in explaining species invasions. *Trends in Ecology and Evolution*, **20**, 223–228.

Lockwood, J. L., Cassey, P., & Blackburn, T. M. (2009) The more you introduce the more you get: the role of colonization pressure and propagule pressure in invasion ecology. *Diversity and Distributions*, **15**, 904–910.

Lockwood, J. L., Simberloff, D., McKinney, M. L., & Von Holle, B. (2001) How many, and which, plants will invade natural areas? *Biological Invasions*, **3**, 1–8.

Lombaert, E., Guillemaud, T., Cornuet, J.-M., et al. (2010) Bridgehead effect in the worldwide invasion of the biocontrol Harlequin Ladybird. *PLoS ONE*, **5**, e9743, doi: 9710.1371/journal.pone.0009743.

Long, R. L., Panetta, F. D., Steadman, K. J., et al. (2008) Seed persistence in the field may be predicted by laboratory-controlled aging. *Weed Science*, **56**, 523–528.

Lopian, R. (2005) The International Plant Protection Convention and invasive alien species: identification of risks and management of invasive species using the IPPC framework. *Proceedings of the Workshop on Invasive Alien Species and the International Plant Protection Convention*, 22–26 September, pp. 6–16. IPPC Secretariat, Braunschweig.

Lowe, S., Browne, M., Boudjelas, S., & De Poorter, M. (2000) 100 of the world's worst invasive alien species: a selection from the Global Invasive Species Database. Invasive Species Specialist Group (ISSG), World Conservation Union (IUCN).

Lym, R. G. & Kirby, D. R. (1987) Cattle foraging behavior in leafy spurge (*Euphoria esula*) infested rangeland. *Weed Technology*, **1**, 314–318.

Mack, R. N. & Foster, S. K. (2009) Eradicating plant invaders: combining ecologically-based tactics and broad-sense strategy. *Management of Invasive Weeds* (ed. Inderjit), pp. 35–60. Springer, Berlin.

MacLeod, A., Pautasso, M., Jeger, M. J., & Haines-Young, R. (2010a) Evolution of the international regulation of plant pests and challenges for future plant health. *Food Security*, **2**, 49–70.

MacLeod, A., Pautasso, M., Jeger, M. J., & Haines-Young, R. (2010b) Evolution of the international regulation of plant pests and challenges for plant health. *Food Security*, **2**, 49–70.

Maki, K. & Galatowitsch, S. (2004) Movement of invasive aquatic plants into Minnesota (USA) through horticultural trade. *Biological Conservation*, **118**, 389–396.

Marchante, H. & Marchante, E. (2016) Engaging society to fight invasive alien plants in Portugal: one of the main threats to biodiversity. *Biodiversity and Education for Sustainable Development* (eds P. Castro, U.M. Azeiteiro, P. Bacelar-Nicolau, W. Leal Filho, & A. M. Azul), pp. 107–122. Springer International Publishing, Switzerland.

Marchante, H., Morais, M. C., Gamela, A. & Marchante, E. (in press) Using a WebMapping platform to engage volunteers to collect data on invasive plants distribution. *Transactions in GIS*, doi: 10.1111/tgis.12198.

Maxwell, B. D., Backus, V., Hohmann, M. G., et al. (2012) Comparison of transect-based standard and adaptive sampling methods for invasive plant species. *Invasive Plant Science and Management*, **5**, 178–193.

Maxwell, B. D., Lehnhoff, E., & Rew, L. J. (2009) The rationale for monitoring invasive plant populations as a crucial step for management. *Invasive Plant Science and Management*, **2**, 1–9.

McCanny, S. J. & Cavers, P. B. (1988) Spread of proso millet (*Panicum miliaceum* L) in Ontario, Canada: 2. Dispersal by combines. *Weed Research*, **28**, 67–72.

McCubbins, J. S. N., Endres, A. B., Quinn, L., & Barney, J. N. (2013) Frayed seams in the 'patchwork quilt' of American federalism: an empirical analysis of invasive plant species regulation. *Environmental Law*, **43**, 35–81.

McGeoch, M. A., Butchart, S. H. M., Spear, D., et al. (2010) Global indicators of biological invasion: species numbers, biodiversity impact and policy responses. *Diversity and Distributions*, **16**, 95–108.

McGeoch, M. A., Shaw, J. D., Terauds, A., Lee, J. E., & Chown, S. L. (2015) Monitoring biological invasion across the broader Antarctic: A baseline and indicator framework. *Global Environmental Change*, **32**, 108–125.

McGeoch, M. A., Spear, D., Kleynhans, E. J., & Marais, E. (2012) Uncertainty in invasive alien species listing. *Ecological Applications*, **22**, 959–971.

Menz, K. M., Coote, B. G., & Auld, B. A. (1980) Spatial aspects of weed control. *Agricultural Systems*, **6**, 67–75.

Merow, C., Smith, M. J., Edwards, T. C., et al. (2014) What do we gain from simplicity versus complexity in species distribution models? *Ecography*, **37**, 1267–1281.

Mesgaran, M. B., Cousens, R. D., & Webber, B. L. (2014) Here be dragons: a tool for quantifying novelty due to covariate range and correlation change when projecting species distribution models. *Diversity and Distributions*, **20**, 1147–1159.

Meyerson, L. A. & Reaser, J. K. (2002) Biosecurity: moving towards a comprehensive approach. *Bioscience*, **52**, 593–600.

Michaels, S. (2009) Matching knowledge brokering strategies to environmental policy problems and settings. *Environmental Science and Policy*, **12**, 994–1011.

Miller, J., Murphy, D. J., Brown, G. K., Richardson, D. M., & González-Orozco, C. E. (2011) The evolution and phylogenetic placement of invasive Australian acacias of South Africa. *Diversity and Distributions*, **17**, 848–860.

Minton, M. L. & Mack, R. N. (2010) The naturalization of plant populations: effects of cultivation and population size and density. *Oecologia*, **164**, 399–409.

Mintzberg, H. (1978) Patterns in strategy formation. *Management Science*, **9**, 934–948.

Mintzberg, H. & Waters, J. A. (1985) Of strategies, deliberate and emergent. *Strategic Management Journal*, **6**, 257–272.

Mitchell, R. K., Agle, B. R., & Wood, D. J. (1997) Towards a theory of stakeholder identification and salience: defining the principle of who and what really counts. *The Academy of Management Review*, **22**, 853–886.

Moles, A. T. & Westoby, M. (2006) Seed size and plant strategy across the whole life cycle. *Oikos*, **113**, 91–105.

Moodley, D., Geerts, S., Rebelo, T., Richardson, D. M., & Wilson, J. R. U. (2014) Site-specific conditions influence plant naturalization: the case of alien Proteaceae in South Africa. *Acta Oecologica*, **59**, 62–71.

Moodley, D., Geerts, S., Richardson, D. M., & Wilson, J. R. U. (2013) Different traits determine introduction, naturalization and invasion success in woody plants: Proteaceae as a test case. *PLoS ONE*, **8**, e75078, doi: 75010.71371/journal.pone.0075078.

Moody, M. E. & Mack, R. N. (1988) Controlling the spread of plant invasions: the importance of nascent foci. *Journal of Applied Ecology*, **25**, 1009–1021.

Moore, J. L., Hauser, C. E., Bear, J. L., Williams, N. S. G., & McCarthy, M. A. (2011a) Estimating detection–effort curves for plants using search experiments. *Ecological Applications*, **21**, 601–607.

Moore, J. L., Runge, M. C., Webber, B. L., & Wilson, J. R. U. (2011b) Contain or eradicate? Optimizing the management goal for Australian acacia invasions in the face of uncertainty. *Diversity and Distributions*, **17**, 1047–1059.

Morin, X. & Thuiller, W. (2009) Comparing niche- and process-based models to reduce prediction uncertainty in species range shifts under climate change. *Ecology*, **90**, 1301–1313.

Motloung, R. F., Robertson, M. P., Rouget, M., & Wilson, J. R. U. (2014) Forestry trial data can be used to evaluate climate-based species distribution models in predicting tree invasions. *Neobiota*, **20**, 31–48.

Myers, J. H., Savoie, A., & van Randen, E. (1998) Eradication and pest management. *Annual Review of Entomology*, **43**, 471–491.

National Invasive Species Council (2001) *Meeting the Invasive Species Challenge: National Invasive Species Management Plan*. National Invasive Species Council, Washington, DC.

Navie, S. C., McFadyen, R. E. C., Panetta, F. D., & Adkins, S. W. (1998) *Parthenium hysterophorus* L. The Biology of Australian Weeds, Volume 2 (eds F. D. Panetta, R. H. Groves, & R. C. H. Shepherd), pp. 157–176. RG & FJ Richardson, Melbourne.

Newman, G., Wiggins, A., Crall, A., *et al.* (2012) The future of citizen science: emerging technologies and shifting paradigms. *Frontiers in Ecology and the Environment*, **10**, 298–304.

Nguyen, T. L. T. (2011) The invasive potential of parthenium weed (*Parthenium hysterophorus* L.) in Australia. PhD Thesis, University of Queensland.

Nishida, T., Yamashita, N., Asai, M., *et al.* (2009) Developing a pre-entry weed risk assessment system for use in Japan. *Biological Invasions*, **11**, 1319–1333.

Novoa, A., Kaplan, H., Kumschick, S., Wilson, J. R. U., & Richardson, D. M. (2015a) Soft touch or heavy hand? Legislative approaches for preventing invasions: insights from Cactaceae in South Africa. *Invasive Plant Science and Management*, **8**, 307–316.

Novoa, A., Kaplan, H., Wilson, J. R. U., & Richardson, D. M. (2016) Resolving a prickly situation: involving stakeholders in invasive cactus management in South Africa. *Environmental Management*, **57**, 998–1008.

Novoa, A., Le Roux, J. J., Robertson, M. P., Wilson, J. R. U., & Richardson, D. M. (2015b) Introduced and invasive cactus species: a global review. *AoB Plants*, **7**, doi: 010.1093/aobpla/plu1078.

Nuñez, M. A., Moretti, A., & Simberloff, D. (2011) Propagule pressure hypothesis not supported by an 80-year experiment on woody species invasion. *Oikos*, **120**, 1311–1316.

Odom, D. I. S., Cacho, O. J., Sinden, J. A., & Griffith, G. R. (2003) Policies for the management of weeds in natural ecosystems: the case of scotch broom (*Cytisus scoparius*, L.) in an Australian national park. *Ecological Economics*, **44**, 119–135.

Panetta, F. D. (1993) A system for assessing proposed plant introductions for weed potential. *Plant Protection Quarterly*, **8**, 10–14.

Panetta, F. D. (2004) Seed banks: the bane of the weed eradicator. *Proceedings of the 14th Australian Weeds Conference* (eds B. M. Sindel & S. D. Johnson), pp. 523–526. Weeds Society of New South Wales, Sydney.

Panetta, F. D. (2007) Evaluation of weed eradication programs: containment and extirpation. *Diversity and Distributions*, **13**, 33–41.

Panetta, F. D. (2009) Weed eradication: an economic perspective. *Invasive Plant Science and Management*, **2**, 360–368.

Panetta, F. D. (2012) Evaluating the performance of weed containment programmes. *Diversity and Distributions*, **18**, 1024–1032.

Panetta, F. D. (2015) Weed eradication feasibility: lessons of the 21st century. *Weed Research*, **55**, 226–238.

Panetta, F. D. & Cacho, O. J. (2012) Beyond fecundity control: which weeds are most containable? *Journal of Applied Ecology*, **49**, 311–321.

Panetta, F. D. & Cacho, O. J. (2014) Designing weed containment strategies: an approach based on feasibilities of eradication and containment. *Diversity and Distributions*, **20**, 555–566.

Panetta, F. D., Cacho, O. J., Hester, S. M., & Sims-Chilton, N. M. (2011a) Estimating the duration and cost of weed eradication programmes. *Island Invasives: Eradication and Management* (eds C. R. Veitch, M. N. Clout, & D. R. Towns), pp. 472–476. IUCN, Gland.

Panetta, F. D., Cacho, O., Hester, S., Sims-Chilton, N., & Brooks, S. (2011b) Estimating and influencing the duration of weed eradication programmes. *Journal of Applied Ecology*, **48**, 980–988.

Panetta, F. D., Csurhes, S., Markula, A., & Hannan-Jones, M. (2011c) Predicting the cost of eradication for 41 Class 1 declared weeds in Queensland. *Plant Protection Quarterly*, **26**, 42–46.

Panetta, F. D. & Lawes, R. (2005) Evaluation of weed eradication programs: the delimitation of extent. *Diversity and Distributions*, **11**, 435–442.

Panetta, F. D. & Lawes, R. (2007) Evaluation of the Australian branched broomrape (*Orobanche ramosa*) eradication program. *Weed Science*, **55**, 644–651.

Panetta, F. D. & Timmins, S. M. (2004) Evaluating the feasibility of eradication for terrestrial weed invasions. *Plant Protection Quarterly*, **19**, 5–11.

Parker, I. M., Simberloff, D., Lonsdale, W. M., et al. (1999) Impact: toward a framework for understanding the ecological effects of invaders. *Biological Invasions*, **1**, 3–19.

Parkes, J. P. (2006) Eradication of vertebrate pests: are there any general lessons? *Advances in Vertebrate Pest Management IV* (eds C. J. Feare & D. P. Cowan), pp. 91–110. Filander Verlag, Furth.

Parkes, J. P. & Panetta, F. D. (2009) Eradication of pests and weeds: progress and emerging issues in the 21st century. *Invasive Species Management: A Handbook of Techniques* (eds M. N. Clout & P. A. Williams), pp. 47–60. Oxford University Press, Oxford.

Paynter, Q., Csurhes, S. M., Heard, T. A., et al. (2003) Worth the risk? Introduction of legumes can cause more harm than good: an Australian perspective. *Australian Systematic Botany*, **16**, 81–88.

Pemberton, R. W. & Liu, H. (2009) Marketing time predicts naturalization of horticultural plants. *Ecology*, **90**, 69–80.

Perleberg, D. (1999) *Evaluation of Aquatic Plant Trade in Minnesota*. Minnesota Department of Natural Resources Ecological Services, St. Paul, MN.

Pheloung, P. C. (2001) Weed risk assessment of plant introductions to Australia. *Weed Risk Assessment* (eds R. H. Groves, F. D. Panetta, & J. G. Virtue), pp. 83–92. CSIRO, Melbourne.

Pheloung, P. C., Williams, P. A., & Halloy, S. R. (1999) A weed risk assessment model for use as a biosecurity tool evaluating plant introductions. *Journal of Environmental Management*, **57**, 239–251.

Phillips, C., Brown, K., Green, C., *et al.* (2015) *Pieris brassicae* (Great White Butterfly) *Eradication Annual Report 2014/15*. Department of Conservation, Wellington.

Pimentel, D., McNair, S., Janecka, J., *et al.* (2001) Economic and environmental threats of alien plant, animal, and microbe invasions. *Agriculture, Ecosystems & Environment*, **84**, 1–20.

Pluess, T., Cannon, R., Jarošík, V., *et al.* (2012a) When are eradication campaigns successful? A test of common assumptions. *Biological Invasions*, **14**, 1365–1378.

Pluess, T., Jarošík, V., Pyšek, P., *et al.* (2012b) Which factors affect the success or failure of eradication campaigns against alien species? *PLoS ONE*, **7**, e48157, doi: 48110.41371/journal.pone.0048157.

Prentis, P. J., Wilson, J. R. U., Dormontt, E. E., Richardson, D. M., & Lowe, A. J. (2008) Adaptive evolution in invasive species. *Trends in Plant Science*, **13**, 288–294.

Prider, J., Correll, R., & Warren, P. (2012) A model for risk-based assessment of *Phelipanche mutelii* (branched broomrape) eradication in fields. *Weed Research*, **52**, 526–534.

Proches, Ş., Wilson, J. R. U., Richardson, D. M., & Rejmánek, M. (2008) Searching for phylogenetic pattern in biological invasions. *Global Ecology and Biogeography*, **17**, 5–10.

Prosser, C. W., Anderson, G. L., Wendel, L. E., Richard, R. D., & Redlin, B. R. (2002) TEAM leafy spurge: an areawide pest management program. *Integrated Pest Management Reviews*, **7**, 47–62.

Pyšek, P. (1998) Is there a taxonomic pattern to plant invasions? *Oikos*, **82**, 282–294.

Pyšek, P. & Hulme, P. E. (2005) Spatio-temporal dynamics of plant invasions: linking pattern to process. *Ecoscience*, **12**, 302–315.

Pyšek, P., Hulme, P. E., Meyerson, L. A., *et al.* (2013) Hitting the right target: taxonomic challenges for, and of, plant invasions. *AoB Plants*, **5**, doi: 010.1093/aobpla/plt1042.

Pyšek, P., Jarošík, V., Hulme, P. E., *et al.* (2012) A global assessment of invasive plant impacts on resident species, communities and ecosystems: the interaction of impact measures, invading species' traits and environment. *Global Change Biology*, **18**, 1725–1737.

Pyšek, P., Jarošík, V., & Pergl, J. (2011) Alien plants introduced by different pathways differ in invasion success: unintentional introductions as a threat to natural areas. *PLoS ONE*, **6**, e24890, doi: 24810.21371/journal.pone.0024890.

Pyšek, P. & Prach, K. (1993) Plant invasions and the role of riparian habitats: a comparison of 4 species alien to central Europe. *Journal of Biogeography*, **20**, 413–420.

Pyšek, P. & Richardson, D. M. (2007) Traits associated with invasiveness in alien plants: where do we stand? *Biological Invasions* (ed. W. Nentwig), pp. 97–125. Springer-Verlag, Berlin.

Quinn, L. D. (2014) What would invasive feedstock populations look like? Perspectives from existing invasions. *Bioenergy and Biological Invasions: Ecological, Agronomic and Policy Perspectives on Minimising Risk* (eds L. D. Quinn, D. P. Matlaga, & J. N. Barney), pp. 12–34. CABI, Wallingford.

Quinn, L. D., Barney, J. N., McCubbins, J. S. N., & Endres, A. B. (2013) Navigating the 'noxious' and 'invasive' regulatory landscape: suggestion for improved regulation. *Bioscience*, **63**, 124–131.

Quinn, L. D., Gordon, D. R., Glaser, A., Lieurance, D., & Flory, S. L. (2014a) Bioenergy feedstocks at low risk for invasion in the U.S.: a 'white list' approach. *Bioenergy Research*, doi: 10.1007/s12155-014-9503-z.

Quinn, L. D., Scott, E. C., Endres, A. B., *et al.* (2014b) Resolving regulatory uncertainty: legislative language for potentially invasive bioenergy feedstocks. *GCB Bioenergy*, doi: 10.1111/gcbb.12216.

Raghu, S., Anderson, R. C., Daehler, C. C., *et al.* (2006) Adding biofuels to the invasive species fire? *Science*, **313**, 1742–1742.

Randall, R. P. (2012) *A Global Compendium of Weeds*, 2nd edn. Department of Agriculture and Food, Western Australia.

Reed, M. S., Graves, A., Dandy, N., *et al.* (2009) Who's in and why? A typology of stakeholder analysis methods for natural resource management. *Journal of Environmental Management*, **90**, 1933–1949.

Regan, T. J., McCarthy, M. A., Baxter, P. W. J., Panetta, F. D., & Possingham, H. P. (2006) Optimal eradication: when to stop looking for an invasive plant. *Ecology Letters*, **9**, 759–766.

Reichard, S., Schmitz, D. C., Simberloff, D., *et al.* (2005) The tragedy of the commons revisited: invasive species. *Frontiers in Ecology and the Environment*, **3**, 109–115.

Rejmánek, M. (2011) Invasiveness. *Encyclopedia of Biological Invasions* (eds D. Simberloff & M. Rejmánek), pp. 379–385. University of California Press, Berkeley and Los Angeles.

Rejmánek, M. & Pitcairn, M. J. (2002) When is eradication of exotic pest plants a realistic goal? *Turning the Tide: The Eradication of Invasive Species* (eds C. R. Veitch & M. N. Clout), pp. 249–253. IUCN SSC Invasive Species Specialist Group, IUCN, Gland.

Rejmánek, M. & Richardson, D. M. (1996) What attributes make some plant species more invasive? *Ecology*, **77**, 1655–1661.

Rejmánek, M., Richardson, D. M., Higgins, S. I., Pitcairn, M., & Grotkopp, E. (2005) Ecology of invasive plants: state of the art. *Invasive Alien Species: A New Synthesis* (eds H. A. Mooney, R. N. Mack, J. A. McNeely, *et al.*), pp. 104–161. Island Press, Washington, DC.

Renner, I. W., Elith, J., Baddeley, A., *et al.* (2015) Point process models for presence-only analysis. *Methods in Ecology and Evolution*, **6**, 366–379.

Rentería, J. L., Gardener, M. R., Panetta, F. D., & Crawley, M. J. (2012) Management of the invasive hill raspberry (*Rubus niveus*) on Santiago Island, Galapagos: eradication or indefinite control? *Invasive Plant Science and Management*, **5**, 37–46.

Ricciardi, A. (2007) Are modern biological invasions an unprecedented form of global change? *Conservation Biology*, **21**, 329–336.

Ricciardi, A. & Cohen, J. (2007) The invasiveness of an introduced species does not predict its impact. *Biological Invasions*, **9**, 309–315.

Ricciardi, A., Hoopes, M. F., Marchetti, M. P., & Lockwood, J. L. (2013) Progress toward understanding the ecological impacts of nonnative species. *Ecological Monographs*, **83**, 263–282.

Richardson, D. M. (2011) Invasion science: the roads travelled and the roads ahead. *Fifty Years of Invasion Ecology: The Legacy of Charles Elton*, (ed. D. M. Richardson), pp. 397–408. Wiley-Blackwell, Malden, MA.

Richardson, D. M., Allsopp, N., D'Antonio, C. M., Milton, S. J., & Rejmánek, M. (2000a) Plant invasions: the role of mutualisms. *Biological Reviews*, **75**, 65–93.

Richardson, D. M., Carruthers, J., Hui, C., et al. (2011) Human-mediated introductions of Australian acacias: a global experiment in biogeography. *Diversity and Distributions*, **17**, 771–787.

Richardson, D. M., Cowling, R. M., & Le Maitre, D. C. (1990) Assessing the risk of invasive success in *Pinus* and *Banksia* in South African Mountain Fynbos. *Journal of Vegetation Science*, **1**, 629–642.

Richardson, D. M., Le Roux, J. J., & Wilson, J. R. U. (2015) Australian acacias as invasive species: lessons to be learnt from regions with long planting histories. *Southern Forests*, **77**, 31–39.

Richardson, D. M., Pyšek, P., Rejmánek, M., et al. (2000b) Naturalization and invasion of alien plants: concepts and definitions. *Diversity and Distributions*, **6**, 93–107.

Richardson, D. M., van Wilgen, B. W., & Nunez, M. A. (2008) Alien conifer invasions in South America: short fuse burning? *Biological Invasions*, **10**, 573–577.

Richter, R., Dullinger, S., Essl, F., Leitner, M., & Vogl, G. (2013) How to account for habitat suitability in weed management programmes? *Biological Invasions*, **15**, 657–669.

Robertson, M. P., Cumming, G. S., & Erasmus, B. F. N. (2010) Getting the most out of atlas data. *Diversity and Distributions*, **16**, 363–375.

Robertson, M. P., Visser, V. and Hui, C. (2016). Biogeo: an R package for assessing and improving data quality of occurrence record datasets. *Ecography*, **39**, doi: 10.1111/ecog.02118.

Rouget, M., Hui, C., Rentería, J., Richardson, D. M., & Wilson, J. R. U. (2015) Plant invasions as a biogeographical assay: vegetation biomes constrain the distribution of invasive alien species assemblages. *South African Journal of Botany*, **101**, 24–31.

Rouget, M., Robertson, M. P., Wilson, J. R. U., et al. (2016) Invasion debt: quantifying future biological invasions. *Diversity and Distributions*, **22**, 445–456.

Roura-Pascual, N., Richardson, D. M., Krug, R. M., et al. (2009) Ecology and management of alien plant invasions in South African fynbos: accommodating key complexities in objective decision making. *Biological Conservation*, **142**, 1595–1604.

Rout, T. M., Hauser, C. E., & Possingham, H. P. (2009) Optimal adaptive management for the translocation of a threatened species. *Ecological Applications*, **19**, 515–526.

Rout, T. M., Salomon, Y., & McCarthy, M. A. (2009) Using sighting records to declare eradication of an invasive species. *Journal of Applied Ecology*, **46**, 110–117.

Sakai, A. K., Allendorf, F. W., Holt, J. S., et al. (2001) The population biology of invasive species. *Annual Review of Ecology and Systematics*, **32**, 305–332.

Schoeman, J., Buckley, Y. M., Cherry, H., Long, R. L., & Steadman, K. J. (2010) Inter-population variation in seed longevity for two invasive weeds: *Chrysanthemoides monilifera* ssp. *monilifera* (boneseed) and ssp. *rotundata* (bitou bush). *Weed Research*, **50**, 67–75.

Schreck Reis, C., Marchante, H., Freitas, H., & Marchante, E. (2013) Public perception of invasive plant species: assessing the impact of workshop activities to promote young students' awareness. *International Journal of Science Education*, **35**, 690–712.

Scott, J. K. & Batchelor, K. L. (2014) Management of *Chrysanthemoides monilifera* subsp *rotundata* in Western Australia. *Invasive Plant Science and Management*, **7**, 190–196.

Segerson, K. & Miceli, T. J. (1998) Voluntary environmental agreements: good or bad news for environmental protection? *Journal of Environmental Economics and Management*, **36**, 109–130.

Sharov, A. A. & Liebhold, A. M. (1998) Bioeconomics of managing the spread of exotic pest species with barrier zones. *Ecological Applications*, **8**, 833–845.

Shaw, J. D. (2014) Southern Ocean islands as protected areas. *Plant Invasions in Protected Areas* (eds L. Foxcroft, D. M. Richardson, P. Pyšek, & P. Genovesi), pp. 449–470. Springer, Dordrecht.

Shaw, J. D., Wilson, J. R. U., & Richardson, D. M. (2010) Initiating dialogue between scientists and managers of biological invasions. *Biological Invasions*, **12**, 4077–4083.

Shaw, R., Parr, M., Pollard, K., & Williams, F. (2014) Demonstrating the cost of invasive species to Great Britain. *CABI Impact Case Study Series*, **1**, doi: 10.1079/CABICOMM-64–54.

Shefferson, R. P. (2009) The evolutionary ecology of vegetative dormancy in mature herbaceous perennial plants. *Journal of Ecology*, **97**, 1000–1009.

Shimono, Y. & Konuma, A. (2008) Effects of human-mediated processes on weed species composition in internationally traded grain commodities. *Weed Research*, **48**, 10–18.

Shine, C. (2007) Invasive species in an international context: IPPC, CBD, European Strategy on Invasive Alien Species and other legal instruments. *EPPO Bulletin*, **37**, 103–113.

Shine, C. N., Williams, N., & Burhenne-Guilmin, F. (2005) Legal and institutional frameworks for invasive alien species. *Invasive Alien Species: A New Synthesis* (eds H. Mooney, R. Mack, J. McNelly, et al.), pp. 233–284. Island Press, Washington DC.

Shipley, R. (1992) Visioning in planning: is the practice based on sound theory? *Environment and Planning*, 34, 7–22.

Siebert, S., Doll, P., Hoogeveen, J., et al. (2005) Development and validation of the global map of irrigation areas. *Hydrology and Earth System Sciences*, **9**, 535–547.

Silvertown, J., Harvey, M., Greenwood, R., et al. (2015) Crowdsourcing the identification of organisms: a case-study of iSpot. *Zookeys*, **480**, 125–146.

Simberloff, D. (2002) Today Tiritiri Matangi, tomorrow the world! Are we aiming too low in invasives control? *Turning the Tide: The Eradication of Island Invasives* (eds C. R. Vietch & M. N. Clout), pp. 4–12. IUCN, Gland.

Simberloff, D. (2003) Eradication: preventing invasions at the outset. *Weed Science*, **51**, 247–253.

Simberloff, D. (2009) We can eliminate invasions or live with them: successful management projects. *Biological Invasions*, **11**, 149–157.

Simberloff, D. (2013) Eradication: pipe dream or real option? *Plant Invasions in Protected Areas: Patterns, Problems and Challenges* (eds L. C. Foxcroft, P. Pyšek, D. M. Richardson, & P. Genovesi), pp. 549–559. Springer, New York.

Simberloff, D., Nunez, M. A., Ledgard, N. J., *et al.* (2010) Spread and impact of introduced conifers in South America: lessons from other southern hemisphere regions. *Austral Ecology*, **35**, 489–504.

Smith, R. I. L. & Richardson, M. (2011) Fuegian plants in Antarctica: natural or anthropogenically assisted immigrants? *Biological Invasions*, **13**, 1–5.

Smolik, M. G., Dullinger, S., Essl, F., *et al.* (2010) Integrating species distribution models and interacting particle systems to predict the spread of an invasive alien plant. *Journal of Biogeography*, **37**, 411–422.

Sorda, G., Banse, M., & Kemfert, C. (2010) An overview of biofuel policies around the world. *Energy Policy*, **38**, 6977–6988.

Spring, D. & Cacho, O. J. (2015) Estimating eradication probabilities and trade-offs for decision analysis in invasive species eradication programs. *Biological Invasions*, **17**, 191–204.

Sutherland, W. J., Fleishman, E., Mascia, M. B., Pretty, J., & Rudd, M. A. (2011) Methods for collaboratively identifying research priorities and emerging issues in science and policy. *Methods in Ecology and Evolution*, **2**, 238–247.

Sutherst, R. W. & Maywald, G. F. (1985) A computerised system for matching climates in ecology. *Agriculture, Ecosystems & Environment*, **13**, 281–299.

Tasker, A. V. & Westwood, J. H. (2012) The U.S. Witchweed eradication effort turns 50: a retrospective and look-ahead on parasitic weed management. *Weed Science*, **60**, 267–268.

Taylor, C. M. & Hastings, A. (2004) Finding optimal control strategies for invasive species: a density-structured model for *Spartina alterniflora*. *Journal of Applied Ecology*, **41**, 1049–1057.

Terauds, A., Chown, S. L., Morgan, F., *et al.* (2012) Conservation biogeography of Antarctica. *Diversity and Distributions*, **18**, 726–741.

Thompson, K., Jalili, A., Hodgson, J. G., *et al.* (2001) Seed size, shape and persistence in the soil in an Iranian flora. *Seed Science Research*, **11**, 345–355.

Thomson, F. J., Moles, A. T., Auld, T. D., & Kingsford, R. T. (2011) Seed dispersal distance is more strongly correlated with plant height than with seed mass. *Journal of Ecology*, **99**, 1299–1307.

Thomson, F. J., Moles, A. T., Auld, T. D., *et al.* (2010) Chasing the unknown: predicting seed dispersal mechanisms from plant traits. *Journal of Ecology*, **98**, 1310–1318.

Thum, R. A., Mercer, A. T., & Wcisel, D. J. (2012) Loopholes in the regulation of invasive species: genetic identifications identify mislabeling of prohibited aquarium plants. *Biological Invasions*, **14**, 929–937.

Timmons, F. L. (1970) A history of weed control in the United States and Canada. *Weed Science*, **2**, 294–307.

Trakhtenbrot, A., Nathan, R., Perry, G., & Richardson, D. M. (2005) The importance of long-distance dispersal in biodiversity conservation. *Diversity and Distributions*, **11**, 173–181.

Trueman, M., Atkinson, R., Guezou, A., & Wurm, P. (2010) Residence time and human-mediated propagule pressure at work in the alien flora of Galapagos. *Biological Invasions*, **12**, 3949–3960.

United Nations Environment Programme (2010) COP 10 decision X/2. Strategic plan for biodiversity 2011–2020 and the Aichi biodiversity targets. *Conference of the Parties to the Convention on Biological Diversity*. Tenth meeting, Nagoya, 18–29 October 2010. www.cbd.int/doc/decisions/cop-10/cop-10-dec-02-en.pdf.

Václavík, T. & Meentemeyer, R. K. (2009) Invasive species distribution modeling (iSDM): are absence data and dispersal constraints needed to predict actual distributions? *Ecological Modelling*, **220**, 3248–3258.

Van Driesche, J. & Van Driesche, R. (2004) *Nature Out of Place: Biological Invasion in the Global Age*. Island Press, Washington, DC.

van Kleunen, M., Dawson, W., Essl, F., et al. (2015) Global exchange and accumulation of non-native plants. *Nature*, **525**, 100–103.

van Kleunen, M., Dawson, W., & Maurel, N. (2015) Characteristics of successful alien plants. *Molecular Ecology*, **24**, 1954–1968.

van Kleunen, M., Dawson, W., Schlaepfer, D., Jeschke, J. M., & Fischer, M. (2010) Are invaders different? A conceptual framework of comparative approaches for assessing determinants of invasiveness. *Ecology Letters*, **13**, 947–958.

van Kleunen, M., Weber, E., & Fischer, M. (2010) A meta-analysis of trait differences between invasive and non-invasive plant species. *Ecology Letters*, **13**, 235–245.

van Klinken, R. D., Murray, J. V., & Smith, C. (2015) Process-based pest risk mapping using Bayesian networks. *Pest Risk Modelling for Invasive Alien Species* (ed. R. C. Venette), pp. 170–188. CABI, Wallingford.

van Klinken, R. D., Panetta, F. D., & Coutts, S. R. (2013) Are high-impact species predictable? An analysis of naturalised grasses in northern Australia. *PLoS ONE*, **8**, e68678, doi: 68610.61371/journal.pone.0068678.

van Wilgen, B. W., Dyer, C., Hoffmann, J. H., et al. (2011) National-scale strategic approaches for managing introduced plants: insights from Australian acacias in South Africa. *Diversity and Distributions*, **17**, 1060–1075.

van Wilgen, B. W., Forsyth, G. G., Le Maitre, D. C., et al. (2012) An assessment of the effectiveness of a large, national-scale invasive alien plant control strategy in South Africa. *Biological Conservation*, **148**, 28–38.

van Wilgen, B. W., Le Maitre, D. C., & Cowling, R. M. (1998) Ecosystem services, efficiency, sustainability and equity: South Africa's Working for Water programme. *Trends in Ecology and Evolution*, **13**, 378.

van Wilgen, B. W. & Richardson, D. M. (2014) Challenges and trade-offs in the management of invasive alien trees. *Biological Invasions*, **16**, 721–734.

Veldtman, R., Chown, S. L., & McGeoch, M. A. (2010) Using scale–area curves to quantify the distribution, abundance and range expansion potential of an invasive species. *Diversity and Distributions*, **16**, 159–169.

Venette, R. C., Kriticos, D. J., Magarey, R. D., *et al.* (2010) Pest risk maps for invasive alien species: a roadmap for improvement. *Bioscience*, **60**, 349–362.

Verloove, F. (2010) Invaders in disguise: conservation risks derived from misidentifications of invasive plants. *Management of Biological Invasions*, **1**, 1–5.

Vilà, M., Espinar, J. L., Hejda, M., *et al.* (2011) Ecological impacts of invasive alien plants: a meta-analysis of their effects on species, communities and ecosystems. *Ecology Letters*, **14**, 702–708.

Vilà, M., Rohr, R. P., Espinar, J. L., *et al.* (2015) Explaining the variation in impacts of non-native plants on local-scale species richness: the role of phylogenetic relatedness. *Global Ecology and Biogeography*, **24**, 139–146.

Visser, V., Langdon, B., Pauchard, A., & Richardson, D. M. (2014) Unlocking the potential of Google Earth as a tool in invasion science. *Biological Invasions*, **16**, 513–534.

Vittoz, P. & Engler, R. (2007) Seed dispersal distances: a typology based on dispersal modes and plant traits. *Botanica Helvetica*, **117**, 109–124.

Waage, J. K. & Mumford, J. D. (2007) Agricultural biosecurity. *Philosophical Transactions of the Royal Society of London B: Biological Sciences*, **363**, 863–876.

Ware, C., Bergstrom, D. M., Muller, E., & Alsos, I. G. (2012) Humans introduce viable seeds to the Arctic on footwear. *Biological Invasions*, **14**, 567–577.

Waugh, J. (2009) *Neighborhood Watch: Early Detection and Rapid Response to Biological Invasion Along US Trade Pathways*. IUCN, Gland.

Weber, J., Panetta, F. D., Virtue, J., & Pheloung, P. (2009) An analysis of assessment outcomes from eight years' operation of the Australian border weed risk assessment system. *Journal of Environmental Management*, **90**, 798–807.

Welch, B. A., Geissler, P. H., & Latham, P. (2014) *Early Detection of Invasive Plants: Principles and Practices*. US Department of the Interior, US Geological Survey, Reston, VA.

Williams, F., Eschen, R., Harris, A., *et al.* (2010) *The Economic Cost of Invasive Non-native Species on Great Britain*. CABI, Wallingford.

Williams, P. A., Nicol, E., & Newfield, M. (2001) Assessing the risk to indigenous biota of plant taxa new to New Zealand. *Weed Risk Assessment* (eds R. H. Groves, F. D. Panetta, & J. G. Virtue), pp. 100–116. CSIRO, Melbourne.

Williamson, M., Dehnen-Schmutz, K., Kuhn, I., *et al.* (2009) The distribution of range sizes of native and alien plants in four European countries and the effects of residence time. *Diversity and Distributions*, **15**, 158–166.

Williamson, M., Pyšek, P., Jarošík, V., & Prach, K. (2005) On the rates and patterns of spread of alien plants in the Czech Republic, Britain, and Ireland. *Ecoscience*, **12**, 424–433.

Williamson, M. H. & Fitter, A. (1996) The characters of successful invaders. *Biological Conservation*, **78**, 163–170.

Wilson, J. R. U., Caplat, P., Dickie, I., *et al.* (2014) A standardized set of metrics to assess and monitor tree invasions. *Biological Invasions*, **16**, 535–551.

Wilson, J. R. U., Dormontt, E. E., Prentis, P. J., Lowe, A. J., & Richardson, D. M. (2009) Something in the way you move: dispersal pathways affect invasion success. *Trends in Ecology & Evolution*, **24**, 136–144.

Wilson, J. R. U., Gairifo, C., Gibson, M. R., *et al.* (2011) Risk assessment, eradication, and biological control: global efforts to limit Australian acacia invasions. *Diversity and Distributions*, **17**, 1030–1046.

Wilson, J. R. U., Ivey, P., Manyama, P., & Nänni, I. (2013) A new national unit for invasive species detection, assessment and eradication planning. *South African Journal of Science*, **109** (5/6), doi: 10.1590/sajs.2013/20120111.

Wilson, J. R. U., Richardson, D. M., Rouget, M., et al. (2007) Residence time and potential range: crucial considerations in modelling plant invasions. *Diversity and Distributions*, **13**, 11–22.

Wilson, R. J., Thomas, C. D., Fox, R., Roy, D. B., & Kunin, W. E. (2004) Spatial patterns in species distributions reveal biodiversity change. *Nature*, **432**, 393–396.

Wittenberg, R. & Cock, M. J. W. (2001) *Invasive Alien Species: A Toolkit of Best Prevention and Management Practices*. CABI, Wallingford.

Woldendorp, G. & Bomford, M. (2004) *Weed Eradication: Strategies, Timeframes and Costs*. Bureau of Resource Sciences, Canberra.

Woodford, D., MacIsaac, H., Richardson, D. M., et al. (in review) Confronting the wicked problem of managing invasive species.

Wotton, D. M. & Hewitt, C. L. (2004) Marine biosecurity post-border management: developing incursion response systems for New Zealand. *New Zealand Journal of Marine and Freshwater Research*, **38**, 553–559.

Yamoah, E., Gill, G. S. C., & Massey, E. (2013) Eradication programme for four noxious weeds in New Zealand. *New Zealand Plant Protection*, **66**, 40–44.

Young, K. E. & Schrader, T. S. (2014) Predicting risk of invasive species occurrence: remote-sensing strategies. *Early Detection of Invasive Plants: Principles and Practices* (eds B. A. Welch, P. H. Geissler, & P. Latham), pp. 59–77. US Department of the Interior, US Geological Survey, Reston, VA.

Zadek, S. (2006) The logic of collaborative governance: cooperate responsibility, accountability and the social contract. Paper 17, The Corporate Responsibility Initiative. John F. Kennedy School of Government, Harvard University.

Zenni, R. D. & Nuñez, M. A. (2013) The elephant in the room: the role of failed invasions in understanding invasion biology. *Oikos*, **122**, 801–815.

Zurell, D., Elith, J., & Schroeder, B. (2012) Predicting to new environments: tools for visualizing model behaviour and impacts on mapped distributions. *Diversity and Distributions*, **18**, 628–634.

Zwaenepoel, A., Roovers, P., & Hermy, M. (2006) Motor vehicles as vectors of plant species from road verges in a suburban environment. *Basic and Applied Ecology*, **7**, 83–93.

Index

action plan, 173, 174, 183, 184, 185, 191, 193, 200, 225, 226, 228, 236
active surveillance. *See* surveillance: active
agriculture, 9, 11, 19, 21, 24, 26, 28, 29, 44, 45, 47, 54, 56, 61, 65, 103, 123, 139, 145, 148, 150, 151, 152, 153, 155, 158, 161, 163, 169, 177, 178, 188, 199, 202, 203, 206, 207, 209, 210
aquatic, 1, 26, 38, 44, 68, 75, 105, 145, 160, 173, 178
Australian acacias, 1, 19, 22, 23, 31, 58, 69, 77, 105, 118
Australian Weed Risk Assessment (AWRA). *See* risk
awareness, 3, 66, 67, 84, 85, 143, 144, 157, 160, 164, 168, 171, 175, 176, 181, 187, 191, 203, 210–211, 215, 216, 221, 226

bioeconomic, 102, 105, 110, 186, 224
biofuel, 53, 160, 161–164, 221
biosecurity, 3, 44, 55, 68, 84, 140, 142, 143, 145, 148, 149, 150–153, 154, 158, 167, 169, 174, 200, 222, 224, 229
budget, 3, 9, 57, 67, 84, 96, 97, 102, 104, 107, 108, 109, 110, 131–134, 135, 166–167, 168, 169, 171, 173, 176, 177, 182, 184–189, 191, 192, 195, 204, 205, 206, 208, 210, 215, 216, 224, 225, 226

Cactaceae, 23, 92
Campuloclinium macrocephalum, 1, 3, 4, 50, 81, 226
Caulerpa taxifolia, 1, 11, 50, 75, 200
champion. *See* leadership
Chromolaena odorata, 96, 107, 119, 120, 186, 195
citizen science, 62–67, 203, 212, 221
codes of conduct, 139, 155–158

compliance, 127, 139, 140, 156, 168, 176, 198–199
conflict species, 160–166, 167, 168, 229
consultation, 157, 158, 167, 168, 171, 177, 178–182, 184, 185, 192, 197, 215
containment
   absolute, 84, 88, 125
   area, 99, 125, 126, 127, 128, 230
   barrier zone, 77, 100, 102, 126, 127, 195, 229
   feasibility, xiii, 13, 60, 82, 88, 94, 95, 98, 99–102, 103, 104, 106, 107, 226, 230
   impedance, 100, 101, 106
Convention on Biological Diversity, 3, 140, 146–147, 170, 233
coordinated control, xvii, 3, 81, 82, 84–87, 88, 135, 161, 167, 193, 224, 226, 230
coordination and coordinators, 12, 63, 82, 139, 171, 197, 198, 201–204, 215, 216, 227
costs, 4, 7, 12, 13, 25, 28, 43, 45, 63, 64, 65, 71, 78, 90, 94–99, 100–103, 105, 110, 111, 113, 131–134, 135, 137, 138, 156, 160, 163, 164, 167, 168, 174, 185, 187, 188–189, 199, 207, 208
   cost share, 9, 11, 164, 198, 207
   cost–benefit, 7, 50, 82, 85, 105, 110, 168, 195, 223
   opportunity cost, 108, 134, 223

Darwin's naturalisation hypothesis, 22
decision support models, 3, 110, 186, 197, 202, 224
delimitation, 9, 12, 67, 75–78, 84, 97, 118, 129, 130, 131, 184, 194, 195, 221, 222, 230
detection
   rate, 65, 68, 71–74, 79, 117, 222
   threshold, 71–74, 115, 230
   time to, 70, 73, 74

dispersal
  human-mediated, 14, 26, 46, 47, 48, 49, 61, 92, 94, 99, 101, 127, 136, 141
  long-distance, 33, 46, 47, 48, 61, 94, 127, 221
  pathways, 4, 11, 20, 48, 51, 52, 53–55, 56, 57, 61, 63, 76, 79, 92, 99, 101, 110, 127, 140, 141–142, 144, 147, 148, 158, 170, 175, 209, 210, 217, 219, 221, 234
  vectors, 34, 46, 47, 48, 49, 53–55, 127, 136, 141–142, 236

Early Detection and Rapid Response (EDRR), 11, 144, 147, 155, 183, 187, 201, 206–210, 226, 230
e-commerce, 159–160, 168, 221
eradication
  feasibility, xiii, 13, 60, 77, 82, 83, 88–95, 98, 99, 100, 102, 103, 104, 105, 106, 107, 108, 116, 117, 120, 121, 222, 226, 231
  impedance, 88, 89, 91, 92, 93, 100, 102, 106, 231
eradographs, 118, 129–131, 137, 138, 184, 223
exit strategy, 105, 107, 138, 211

Fallopia spp., 33, 188, 189, 191
farming. *See* agriculture
finance and financing. *See* budget
fire, 19, 26, 28, 37, 80, 162, 195, 200, 208
forestry, 8, 23, 25, 29, 44, 56, 57, 58, 65, 77, 78, 116, 139, 151, 152, 160, 162, 177, 178, 188
freshwater. *See* aquatic
funding. *See* budget

horticulture, 1, 9, 23, 25, 26, 36, 46, 47, 54, 58, 151, 157, 160, 173, 188, 205, 208, 212

International Plant Protection Convention (IPPC), 3, 18, 135, 140, 145–146, 147, 151, 154, 155, 202
International Standards for Phytosanitary Measures, 18, 145, 146, 147, 155, 195
internet, 103, 159–160, 168, 188, 191, 192, 201, 206, 212, 216
introduction dynamics, 13, 20, 36, 37, 48, 50, 53
invasion
  debt, 13–16, 218, 232
  hotspot, 55–57, 63, 64, 78, 79, 221

invasiveness, 13, 19, 20–25, 27, 32, 37, 46, 48, 50, 51, 81, 90, 109, 156, 161, 162, 163, 164, 186, 217

juvenile period, 13, 21, 27, 92, 94, 95, 98, 100, 101, 103

knotweeds. *See Fallopia* spp.

lag phase, 2, 32, 33–36, 233
leadership, 16, 152, 170, 176, 182–183, 187, 191, 194–196, 198, 199, 205, 214, 215, 225, 226
legislation, 54, 68
lists, 8, 27, 29, 51, 67–69, 84, 94, 110, 140, 150, 151, 156, 157, 160, 162, 163, 171, 201, 224
  permitted, 16, 154, 164, 168, 234
  prohibited, 16, 51, 154, 167, 235
  watch, 16, 51, 80, 155, 156, 167, 202, 203, 206, 215

management plan, 87, 142, 148, 166, 177, 211, 233
marine. *See* aquatic
monitoring, 6, 55, 63, 64, 65, 69, 71, 78, 99, 112, 113, 115, 125, 126, 127, 131, 132, 136, 137, 144, 160, 164, 171, 191, 192, 196–198, 203, 205, 226

opportunity cost. *See* costs
ornamental plants. *See* horticulture
Orobanche ramosa, 11, 76, 96, 103, 107, 123, 124, 125, 127, 128, 129, 130, 186

Parthenium hysterophorus, 28, 136, 203
passive surveillance. *See* surveillance: passive
pest management, 145, 152, 206
phenology, 71, 113, 118, 234
pines, 1, 19, 21, 58
plant health, 3, 68, 84, 140, 145, 146, 147, 150–153, 154, 201, 224, 234
pom-pom weed. *See Campuloclinium macrocephalum*
prevention, 2, 4, 14, 15, 16, 17, 55, 81, 84, 139, 140, 145, 147, 148, 150, 151, 152, 158, 161, 164, 169, 170, 171, 174, 175, 186, 187, 201, 202, 203, 207, 209, 221, 224, 225
prioritisation, 3, 4, 7, 22, 32, 43, 44, 64, 74, 82, 107, 108, 109, 132, 133, 138, 147, 161, 171, 173, 178, 179, 181, 183,

188–189, 191, 192, 201, 202, 203, 215, 218, 221, 235
progress score, 119–122, 138
propagule pressure, 21, 23, 25, 37, 46, 54, 142, 144, 217, 235

quarantine, 135, 145, 146, 149, 153, 154, 155, 160, 175, 188, 202, 203, 235

remote sensing, 71, 133, 222
risk
    analysis, 4, 6, 37, 39, 49, 54, 139, 140, 145, 146, 148, 151, 152, 153, 154, 155, 158, 167, 168, 176, 193, 202, 235
    assessment, 4, 22, 29, 32, 36, 37–39, 49, 50, 51, 53, 54, 81, 82, 90, 97, 109, 146, 152, 153, 154, 161, 164, 166, 171, 183, 193, 235
    Australian Weed Risk Assessment (AWRA), 38
    management, 4, 23, 32, 37, 38, 48, 168, 183, 211, 235
    mapping, 40, 44, 49, 57, 58–71, 77, 78, 79, 110, 137, 219

search effort, 69–75, 79, 89, 94, 102, 133, 137, 196, 221
seed
    bank, 7, 15, 53, 71, 85, 86, 92, 95, 97, 98, 100, 103, 105, 108, 110, 117, 118, 119, 121, 125, 129, 135, 137, 185
    persistence, 85, 86, 87, 92, 94, 95, 96, 98, 100, 102, 103, 105, 118, 119, 121, 122, 123, 125, 129, 135, 136, 186, 223
    viability, 36, 87, 117, 118, 205
species distribution models, 40–45, 61, 219, 221, 235
species traits, 13, 19, 20–31, 32, 37, 39, 47, 48, 50, 51, 54, 56, 161, 162, 217
Specific, Measurable, Attainable, Realistic and Time-bound (SMART) goals, 192, 225
spotter networks and schemes, 65, 66, 67, 222, 235
spread rate, 26, 46, 48, 219
SPS Agreement, 140, 146, 147, 159, 236
stakeholders, 27, 38, 79, 81, 85, 99, 107, 109, 152, 154, 155, 156, 160, 166, 167, 168, 169, 171, 175, 176, 177, 178–182, 184, 185, 191, 192, 194, 197, 198, 199, 201, 204, 210, 221, 224, 225
stopping rule, 121, 136–137, 224
Striga asiatica, 11, 103, 207
surveillance, 7, 9, 11, 43, 44, 62–67, 71–74, 79, 85, 88, 131, 132, 133, 145, 151, 155, 184, 185, 188, 194, 195, 196, 203, 207, 221, 226, 236
    active, 62, 63, 64, 66, 67, 77, 78, 79, 101, 132, 133, 137, 183, 236
    passive, 10, 63, 64, 65, 77, 78, 79, 85, 100, 101, 132, 133, 137, 183, 195, 222, 236
switch point, 110, 111, 136, 138

time to maturation. *See* juvenile period

web and websites. *See* internet
workshops, 181, 182, 192, 201, 212, 214, 215